Discovering Imaginary Numbers in Everyday Life

The Popularization of Complex Numbers

Qiujiang Lu

MathWonder

Discovering Imaginary Numbers in Everyday Life

Printed in the United States of America

ISBN: 979-8-9927543-2-2

www.mathwonder.org

To all learners
who keep human curiosity flourishing despite AI

Contents

Introduction

Imaginary numbers, which are real numbers multiplied by the imaginary unit $i = \sqrt{-1}$, were first encountered in the 1520s from solving the cubic equation by Italian mathematician Scipione del Ferro. Complex numbers, each of which is the sum of a real number and an imaginary number, were eventually accepted by Renaissance mathematicians and have deep significance and profound importance to our understanding of mathematics and physics.

Historically, $\sqrt{-1}$ has invoked mystery before the prolonged acceptance of the complex numbers by scholars. The term "imaginary" was coined by French mathematician René Descartes in 1637, who was at pains to stress their unreal nature:

"...sometimes only imaginary, that is one can imagine as many as I said in each equation, but sometimes there exists no quantity that matches that which we imagine."

Even today, the enigma still exists for many beginners who may ask: "Why do we need to study numbers that do not belong to the real world?" This is because there has been no evidence indicating the existence of imaginary numbers in everyday life. From first principles thinking, the contemporary definition $i^2 = -1$ is basically the same as $i = \sqrt{-1}$ despite the symbolic difference. Although the former feels less intimidating, both come to the same result $i^2 = -1$. The property $i^2 = -1$ is used to define i while i itself is also used to define the property. This is a circular argument. The definition $i^2 = -1$ just indicates that i has the property, but i itself is still undefined, or is not based on a first principle or a thing that is already known. A first principle is a fundamental assumption that cannot be further deduced. Ancient Greek philosopher Aristotle (384 - 322 BCE) expressed a first principle as "the first basis from which a thing is known." It is no wonder many students are unsatisfied with $i = \sqrt{-1}$ that originates from nowhere in the real world (the unknown).

There also exist other types of numbers that are more widely understood by people because they have respective physical relevance and connections to real life. Natural numbers are often associated with counting quantities, such as 3 apples. Rational numbers indicate fractions, such as half of an apple. Irrational numbers have infinite and non-repeating decimal expansion. The well-known constant π as the ratio of a circle's circumference to its diameter is an example of an irrational number. Real numbers represent the collection of rational and irrational numbers. Negative numbers express the opposite of a quantity and have practical applications, such as indicating elevation below sea level, while positive numbers denote elevation above it. Positive and negative real numbers can be used to describe properties of the real world. However, for about 500 years since the first encounter with imaginary numbers, their existence in physical reality has remained unknown until recently.

The research work in 2023 by the author has derived the imaginary unit from first principles and revealed the existence of imaginary numbers in our daily life. This opens up the opportunity to explore and share with the general public a newly obtained, comprehensive understanding of the imaginary unit and complex numbers. The work posted on social media has received laudation and contentment from users, which has inspired the writing of the book. Many learners have been unhappy with the trick of $\sqrt{-1}$ that solves a problem but has no real world origin, and could remain pondering for years to come. Further, the unreal state could also discourage the learning and hinder the popularization. The aim of the book is to help general readers "really" understand and get fully settled on the subject. In the process, we may gain more knowledge and appreciation about the wonder and beauty of mathematics and Nature.

1. Yin and yang of number systems

Can we imagine a world without numbers? In the modern era, the development of science, technology, and finance in the advanced society depends on mathematics, which in turn depends on numbers. In ancient times, numbers were initially invented in everyday concerns out of the need for counting. Later, numbers were used for measurement and other purposes with the eventual foundation laying for advanced mathematics concepts, such as geometry. The scopes and concepts of numbers have evolved, and the evolution will continue as science and mathematics development demands the further extension of number concepts.

To understand imaginary and complex numbers as an important and radical number concept evolution step, it is helpful to first see and compare the roles played by various types of numbers that are commonly encountered in our daily life.

1.1 Number systems

Numbers are classified into number sets or number systems. The main number sets with respective representing letters are natural numbers $\mathbb{N}$, integers $\mathbb{Z}$, rational numbers $\mathbb{Q}$, and real numbers $\mathbb{R}$. Each of these number systems is a subset of the next one.

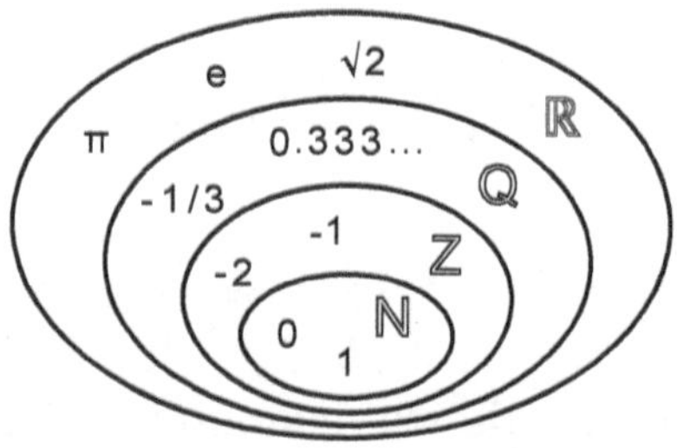

Figure 1.1. Real numbers $\mathbb{R}$ include Rational $\mathbb{Q}$, which include Integer $\mathbb{Z}$, which include Natural numbers $\mathbb{N}$.

For example, an integer is also a rational number, which is also a real number. The belonging relationship can be denoted by

$$\mathbb{N} \in \mathbb{Z} \in \mathbb{Q} \in \mathbb{R}$$

Don't be intimidated by some unfamiliar mathematical symbols. The symbol $\in$ just means "belongs to". For instance, if $A \in B$, it indicates that all members (numbers) in set A are members of set B.

Nowadays, these number systems have been well established. However, historically, when each of the above number systems was introduced, there might have been a hesitance period before the full adoption. For example, the acceptance of negative numbers was resisted by the non-existence of a negative-number quantity of a tangible object, such as "negative-three apples", and negative solutions to problems were regarded "false". The negative number concept was then embraced after its connection to the real world was found through the interpretation of the concept as the opposite of positive amount, such as debts and losses. Establishing the physical world connection for a number system speeds up its adoption process.

In the modern times, there are representations and real-world connections for real numbers and the other number subsets. Revealing the real-world connection for a particular number system invites trust and acceptance, and provides the motivation for people to learn the system.

The imaginary and complex numbers can be considered as the extension of the real number system. To understand the real-world connection and the extension process beyond the real number system, let's first see the complementary and contrary characteristics of the extensions of various number subsets of the real numbers, which may be well described through the concept of yin and yang.

1.2 Yin and yang

Yin and yang, also yinyang or yin-yang, is a fundamental concept that originated in Chinese philosophy and is dated back to the 3rd century BCE or even earlier. Many natural dualities are considered as manifestations of the duality symbolized by yin and yang. This duality lies at the origins of many branches of classical Chinese philosophy, and also serves as a primary guideline of traditional Chinese medicine. An important element of the yin and yang concept is that all things exist as inseparable and contradictory opposites. Examples of yin-yang opposite forces are female-male, dark-light, water-fire, weak-strong, and north-south. The yin-yang concept describes the inherence of duality, paradox, coexistence, interaction, change, unity, and harmony for two mutual opposites. Yin and yang emphasizes the necessity of each to create balance.

Figure 1.2. Yin-yang symbol

The yin-yang symbol contains a circle divided into two halves by a curved line. One half of the circle is black for the yin side while the other is white for the yang. The two opposite sides both attract and complement each other to form a whole. The inverse small dot at each side's core indicates that there is always yin within yang and always yang within yin. The curvy line implies the notion of no absolute separations between the two opposites. The two halves swirl to represent the conversion of yin to yang and yang to yin. The yin-yang symbol shows a balance between two opposites to achieve harmony.

Yin and yang disfavors absolute extremeness, particularly, absolute unchangedness and absolute independence. The only absoluteness in the universe is that nothing is absolute. For example, there is no absolute vacuum anywhere in the universe. Even if we managed to remove any matter (atoms or molecules) from a volume in space, we would find out that it is still full of energy and other things known or unknown to us at this point. There are cosmic microwave background and other types of radiation. And even if we were able to shield all the radiation, we would recognize that there is the presence of gravity waves. And further, even if we could block gravity waves, we would realize that it is full of energy fluctuations and virtual particles as predicted by quantum mechanics. Furthermore, from a different perspective, we would have to use scientific equipment to measure and know if the volume is in a vacuum state. The equipment itself would have to somehow interact with the volume through exchanges of matter, radiation and other things, making it impossible to have the absolute vacuum.

The term yin and yang represents many meanings and principles, and there is no clear-cut definition. In general, the yin-yang philosophy reflects three main characteristics: opposition, interchange and unity. Next we will identify and describe the yin and yang for number subsets in the real number system.

1.3 Zero

Zero is actually a recent development in human history. The first known documented use of zero dates to 628 CE when Indian mathematician Brahmagupta treated 0 as a number and discussed operations about it. The use of 0 as a number should be distinguished from its use as a placeholder numeral such as the one in 103. Our understanding of zero is profound, as we don't encounter zero in the natural world.

Humans didn't understand the number zero for the majority of our history in that we are not born to feel the existence of zero through our sensory systems. Numbers such as one, two, and three have a representative in the physical world. We can see one cat. We can hear the barking of two dogs. But with respect to zero, it requires us to realize that nothing is essentially also a thing by its very nature.

It is noted that in regard to yin and yang, the opposite or the complement part is dependent on the context subject and scope. For instance, with respect to natural numbers, zero may be considered as yin, and nonzero natural numbers as yang. By the same token, with respect to real numbers, zero may be viewed as yin, and nonzero real numbers as yang.

For the scope of natural numbers, the following shows the relevant yin-yang characteristics.

Opposition: A nonzero as yang represents having something, and zero as yin represents the opposite, having nothing.

Interchange: Something and nothing can be converted to each other through operations. For example, number 0 results in 3 by addition of 3, and number 3 results in 0 by subtraction of 3.

Unity: The zero and nonzero complement each other, and form the completeness of the natural number system for description and representation of something and nothing.

The two opposites may not be in a static state, but in change. This is illustrated well in calculus. The notion of limit zero and infinitesimal number is of the essence in calculus. An infinitesimal number is a quantity that is closer to 0 than any standard real number, but that is not 0. The inverse of an infinitesimal is an infinity, and the inverse of an infinity is an infinitesimal, which is close enough to zero to be taken as zero for all practical purposes. The infinitesimal and infinity transform into each other through the inverse. Nothing is everything! The yin-yang philosophy is comfortable with paradox and interchange of the opposites.

1.4 Negative numbers

Nowadays, the concept of negative numbers is very simple and easy to understand even for children. However, as with some other ideas in the history, there had been a long difficult time for its acceptance. In fact, negative numbers didn't become universally adopted until the 19th century.

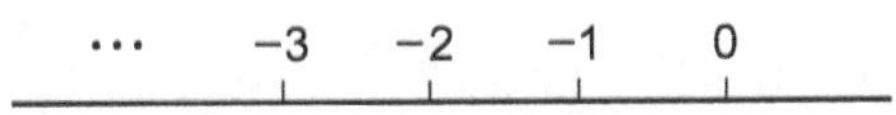

Figure 1.4. A number line with negative numbers to the left of zero

Numbers originated from the need for counting objects in everyday life. Perhaps, the difficulty with understanding negative numbers was related to the counting, which relied on the observation of physical objects. If we see two apples, we can count one or two of them. But, we can't see and count zero apple. Neither can we see and count a negative number of apples.

As early as 100 - 50 BCE in China, the concept of negative numbers was presented in The Nine Chapters on the Mathematical Art. Red rods were used to denote positive coefficients, black for negative. Rules for signed numbers were provided. In the 3rd century CE in Greece, mathematician Diophantus presented the equation equivalent to $4x + 20 = 4$ in his Arithmetica. The equation gave a negative solution, and he called the result "absurd". In the 17th century, English mathematician, John Wallis invented the number line to represent negative numbers along with positive numbers, and described operations through moving forward and backward on the line by addition and subtraction.

For the scope of integer numbers, negative integers may be considered as yin, and positive natural numbers as yang. The relevant yin-yang characteristics are illustrated below.

Opposition: Literally and in concept, negative and positive numbers are the opposite to each other. They are often used to represent the opposite amount, quantity or state. For example, in finance, negative numbers indicate losses in contrast to positive numbers for gains; in temperature measurement on the Celsius scale, negative numbers denote temperatures below zero (indicating water in solid state), while positive numbers express temperatures above zero (indicating water in liquid state); and in tests and games, negative numbers are often used as penalties while positive numbers as scores.

Interchange: Negative and positive numbers can be transformed into each other through operations. For example, a negative number may be obtained by subtraction of two positive numbers, as in $3 - 5 = -2$. And a positive number may be produced by subtraction of two negative numbers, as in $-3 - (-5) = 2$.

Unity: Negative integers and natural numbers complement each other, and form the integer number system with the corresponding extension of the scope.

1.5 Fractional numbers

The word fraction is known to come from the Latin "fractio" which means "a breaking." It represents the concept of breaking a whole into parts. In mathematics, the invention of fractional numbers has reflected the need to express quantities that are parts of a whole such as 1/2 for half of an apple. From as early as 1800 BCE, the Egyptians were writing fractions with a number system in a ten-based notion like ours today. However, the ancient Egyptian writing system then was all in pictures, including those for numbers.

Besides the ratio notation of two whole numbers, fractional numbers are also expressed as decimal fractions. The concept of decimal fractions is intimately connected to decimal place-value representation with the two having been developed alongside each other. The fractional portion of the number is represented by the place values that follow the decimal. The place values after the decimal point start from tenths, followed by hundredths, then thousandths, and so forth.

For the scope of rational numbers, fractional numbers may be considered as yin, and integer numbers as yang. The related yin-yang characteristics are described as follows.

Opposition: In concept, whole and fraction are considered as the opposite to each other. A nonzero integer number is a whole number (or signed whole) and represents a quantity in whole. A fractional number is the opposite, representing a quantity for parts of the whole.

Interchange: Breaking a whole into smaller parts produces fractions. Combining broken parts forms a whole. For example, number 1 can be broken into two equal parts with each part represented by a fractional number 1/2. The whole number 1 can be obtained by combing the parts via addition (1/2 + 1/2) or multiplication (1/2)*2.

Unity: Fractional and integer numbers complement each other, and form the rational number system.

1.6 Irrational numbers

An irrational number is any real number that cannot be expressed as the ratio of two integers. When expressed in positional digit notation, the decimal expansion of an irrational number never repeats and terminates. The first discovered irrational number is probably the square root of 2. For a square with each side having a length of one, the diagonal of the square will be the square root of 2.

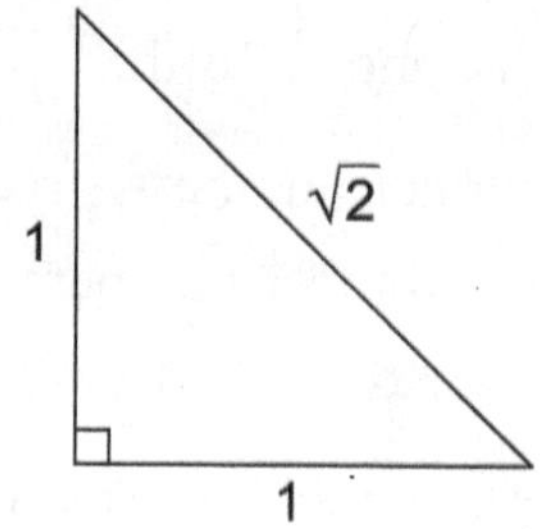

Figure 1.6. Square root of 2

Hippasus (530 - 450 BCE) of Metapontum, a Greek philosopher and early follower of Pythagoras, is usually credited with attaining the first proof of the existence of irrational numbers.

Presumably, Hippasus tried to represent the square root of 2 as a fraction, but failed. He then instead decided to prove that it couldn't be done. According to legend with uncertain ancient reports, the Pythagorean community considered the existence of irrational numbers as a threat to their philosophical worldview and expelled Hippasus from the group. In some versions of the story, he was thrown overboard from a ship.

Unlike fractional and other types of numbers that could be considered as having been invented, the existence of irrational numbers was discovered unintentionally. We can invent certain types of numbers, but once defined, they will follow certain rules that are not invented and need to be discovered. On a separate but related note, this is consistent with the viewpoint that mathematics is both invented and discovered. Mathematics has been developed to represent and describe the reality. It is natural for mathematics to contain both elements of human-originated invention and Nature-embedded discovery. If the two were separated during the development, the mathematics we know today would not exist.

Before Hippasus, it was generally thought that all things in the universe could be measured or counted to by rational numbers. The discovery of irrational numbers might create irritations to some scholars with uncountable infinite long decimal digits, but it is this irrationality or infinity that was later found to be indispensable to the forming of a complete number system for calculus. The number line is filled with infinitely dense rational numbers. But, despite the infinite density, there are still holes in the number line that cannot be represented as a ratio of two integers and need to be filled with irrational numbers.

By the 17th century CE, decimal fractions with modern notation had been generally adopted. However, it was not until the 19th century that mathematicians undertook the thorough study of irrational numbers. Georg Cantor proved that there are more irrational numbers than there are rational numbers, and provided the first rigorous definition of real numbers, which could continuously fill the number line. Contrary to the name, irrational numbers are essential for the formation of real numbers, without which we couldn't do the basic mathematics that makes the world function today.

As for the scope of real numbers, irrational numbers may be viewed as yin, and rational numbers as yang. The pertinent yin-yang characteristics are expressed below.

Opposition: From the perspective of fraction, the irrationality of being unable to be expressed as the ratio of two integers is the opposite to the rationality of being able to. From the perspective of decimal pattern, a rational number has none or a limited number of decimal digits while oppositely an irrational has an infinite number of non-repeating decimals.

Interchange: The interchangeability may be reflected by obtaining one type of number from the opposite type through operations. For example, the square root of rational number 2 is an irrational; and the multiplication of square root of 2 to itself is rational number 2. Also, the interchangeability and inter-containment may be reflected in an infinite process. With regard to the infinitesimal concept in calculus, an infinitesimal number involves infinite numbers of rational and irrational numbers, but the ratio of two infinitesimals can become a single rational or irrational number.

Unity: Irrational and rational numbers complement each other, and form the real number system, which can continuously fill the number line.

1.7 Real numbers

In mathematics, a real number is a value that can be represented by an infinite decimal expansion. Real numbers include rational numbers and irrational numbers. A real number can be used to measure a continuous one-dimensional quantity such as a distance, speed or duration. The adjective real was introduced in the 17th century by René Descartes to distinguish real numbers from imaginary numbers.

Real numbers form a complete unity set, which consists of its subsets. The following table summarizes yin-yang number subsets and their respective unity-sets in the real number system.

Unity set	**Yin subset •**	**Yang subset ○**
Natural ($\mathbb{N}$)	Zero	Nonzero
Integer ($\mathbb{Z}$)	Negative	Natural
Rational ($\mathbb{Q}$)	Fractional	Integer
Real ($\mathbb{R}$)	Irrational	Rational

Table 1.7. Yin-yang number subsets in the real number system

It is the contrary and complement characteristics of the yin-yang subsets that form the complete unity set of real numbers, which in turn form the continuous number line called real number line or real line.

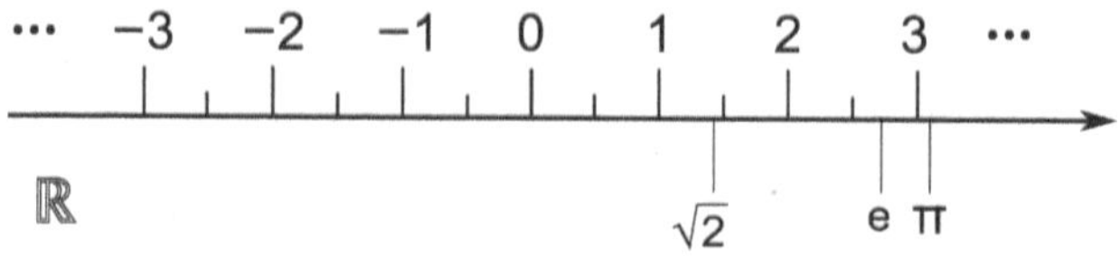

Figure 1.7. Real number line

Here, continuous means that pairs of real number values can have arbitrarily small differences. The continuity in real numbers indicates that divisions of quantity can be infinite. For example, a line segment can be cut into half, which cut into half, which cut into half, and so forth. The process can continue infinitely with the current half becoming closer and closer to zero, but never reaching exactly zero. This means that an infinitely small number can contain an infinite large number of segments, illustrating the concept of yin within yang and vice versa. Real numbers are fundamental for calculus in particular, and for all mathematics in general. The continuous property of the real numbers is the basis on which calculus and broader mathematical analysis are built.

All points in the real number line can be represented by real numbers. Within the scope of one-dimension, the real number system fulfills the continuous number line and is considered complete. In other words, from the dimension perspective, real numbers are one-dimensional numbers.

2. Mathematical representations of points in a plane

Since real numbers are one-dimensional numbers for representing points in a line, it may be natural for one to ask: Are there two-dimensional numbers for representing points in a plane? For the answer to the question, it is worthwhile to first examine what numbers and their extensions actually mean in general, and to investigate how to represent points in a two-dimensional plane in particular.

2.1 Number concept extension beyond one-dimension

In mathematics, the word "number" has no exact definition. As we know, there are four basic arithmetic operations (addition, subtraction, multiplication and division) that can be performed to numbers. And there are four basic properties of numbers (commutative, associative, distributive, and identity) as rules and characteristics that numbers follow when we perform arithmetic operations on them. We apply these properties while doing addition and multiplication operations.

The commutative property of addition and multiplication means that changing the order of the operands does not change the result. Take real numbers for example: $3 + 2 = 2 + 3$ or $3 \times 4 = 4 \times 3$. And the associative property of addition and multiplication means that rearranging the parentheses in an expression will not change the result. Also take real numbers for example: $(2 + 3) + 4 = 2 + (3 + 4)$ or $(2 \times 3) \times 4 = 2 \times (3 \times 4)$. It should be noted that as for the other two basic arithmetic operations, subtraction and division, we can achieve the same effect of subtraction by addition of negative number as in $2–3 = 2+(-3)$, and that of division by multiplication of number's inverse as in $2/3 = 2 \times (1/3)$.

With regard to one-dimension, real numbers represent points in a line, and have the four basic properties along with the four basic operations applicable to them. By analogy, we may then extend the number concept to two-dimensions as long as the new number type possesses all or some of the four basic properties and has the validity for all or some of the four basic operations. The more valid properties and operations the new number type has, the more desirable. Similarly, the number concept may be further extended to higher n-dimensions.

For instance, in a space of higher k-dimensions with k being a particular integer greater than one, if we come up with numbers that represent points in the said space and that have multiplication and addition valid to them, we may then define them as k-dimensional numbers and give them a certain number name. Specifically, assume that numbers, p_1 and p_2 represent two points in the said space. The multiplication $p_1 \cdot p_2$ and addition $p_1 + p_2$ are both valid.

In regard to a number type extension, it will be wonderful if the new numbers also have both or one of the commutative and associative properties. This makes the new numbers less restricted and more useful in arithmetic or algebraic manipulation and analysis. After understanding the number extension concept, we will next focus on two-dimensions and representations of points in a plane.

2.2 Cartesian coordinates

A Cartesian coordinate system is named after French mathematician and philosopher René Descartes, who published the idea in 1637. For two-dimensions, a Cartesian coordinate system in a plane is defined by two perpendicular real number lines: the x-axis and the y-axis. In this coordinate system, we can represent each point in the plane by an ordered pair of real numbers called coordinates, which are the signed distances to the point from x-axis and y-axis, respectively. The point where the two axes meet is called the origin and has coordinates (0, 0).

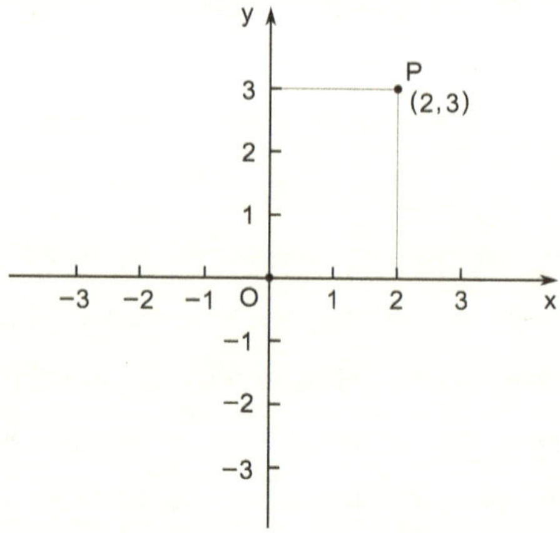

Figure 2.2a. Cartesian coordinate plane with the origin O and a point P at coordinates (2, 3)

The invention of Cartesian coordinates in the 17th century revolutionized mathematics by transforming geometry problems into equations and solving them with algebra and calculus. With a pair of real number variables x and y, we can use x-y coordinates (x, y) to identify an arbitrary point in the plane. For a geometric shape, a set of x-y coordinates corresponding to points of the shape can satisfy an equation that expresses the shape. As an illustration, the equation $x^2 + y^2 = 9$ represents a circle of radius 3, centered at the origin, with the x and y belonging to the set of all points that satisfy the equation.

Basically, in the plane, the Cartesian coordinate system identifies the position of an arbitrary point P relative to the origin O, as illustrated in the figure:

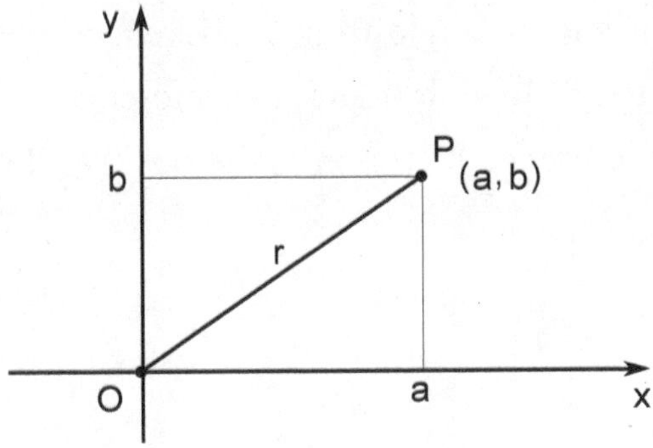

Figure 2.2b. Position point P relative to the origin O

It is important to note that the coordinates (a,b) of the position point P also implicitly represent its direction and distance relative to the origin. The distance of the point P relative to the origin is $r = \sqrt{a^2 + b^2}$, according to Pythagorean theorem. We will next discuss the relative direction of the position point in polar coordinates.

2.3 Polar coordinates

Polar coordinates are particularly effective for analyzing a phenomenon that is inherently associated with direction and length from a center point in a plane. The polar coordinate system was initially introduced for the study of circular and orbital motion. The concepts of the polar system were independently conceived by Grégoire de Saint-Vincent and Bonaventura Cavalieri in the 17th century. However, the actual term "polar coordinates" has been attributed to Gregorio Fontana in the 18th century.

A polar coordinate system is for a two-dimensional coordinate plane where each point is identified by a distance from a reference point and an angle from a reference direction. As an illustration in the following figure, the reference point is the origin O and the reference direction is the positive direction of the x-axis.

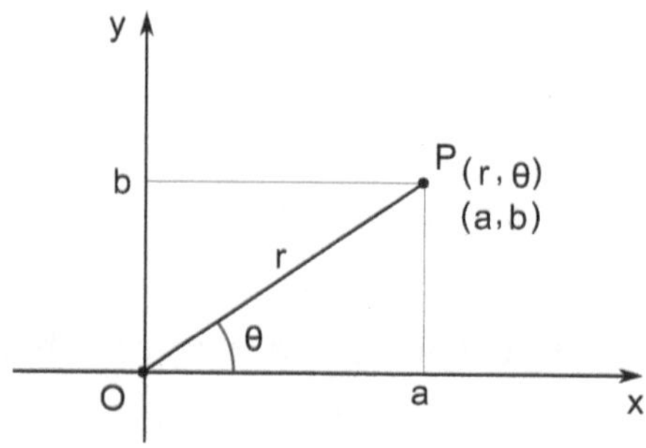

Figure 2.3. Position point P represented by polar coordinates (r, θ) and Cartesian coordinates (a, b)

The position point P is represented by its polar coordinates (r, θ), with r being the radial coordinate for the distance from the origin reference point, and θ being the angular coordinate for the angle from the x-axis reference direction. The Cartesian coordinates (a, b) of the point P are also presented in the figure for comparison.

Angles in the polar system are generally expressed in either radians or degrees. A full turn corresponding to 2π radians or 360° makes the direction return to the same initial direction. Therefore, the polar coordinates (r, θ) represent the same point as coordinates (r, $\theta + \mathrm{n} \cdot 2\pi$), where n is an arbitrary integer. For the unique representation, it is customary to limit θ to either the interval $[0, 2\pi)$ or $(-\pi, \pi]$. The positive θ means counter-clockwise rotation and negative θ clockwise.

As will be seen in due course, the representations for the position of a point in a plane are closely related to the subject of two-dimensional numbers. In addition to Cartesian and polar coordinates, vectors can also be used to represent the position of a point.

2.4 Position vectors

In mathematics and physics, a vector represents a quantity that has both magnitude and direction. Magnitude refers to the length of the vector. Vectors are fundamental in the physical sciences. The displacement, velocity and acceleration of a moving object as well as the forces acting on it can be expressed with vectors. Other physical vectors, such as the electric and magnetic field, are represented as a vector field for each point in a physical space. The mathematical representation of a physical vector depends on the coordinate system used to describe it. A vector is denoted by a line having an arrow, with the length of the line indicating the magnitude of the vector and the arrow showing the direction. By convention, a vector is commonly designated in boldface.

The Latin word vector means "carrier". A vector is intended to "carry" the point A to the point B. It was first used by 18th century astronomers who studied planetary revolution around the Sun.

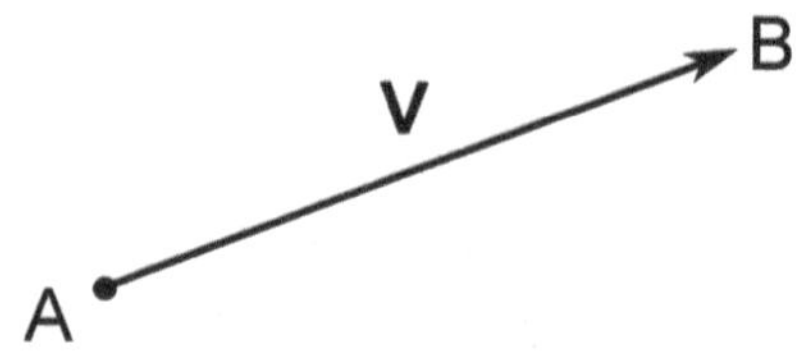

Figure 2.4a. Vector **V** pointing from point A to point B

The magnitude of the vector is the distance between the two points, and the direction indicates the direction of displacement from A to B.

The position of a point in a plane can be represented with a position vector as illustrated in the figure:

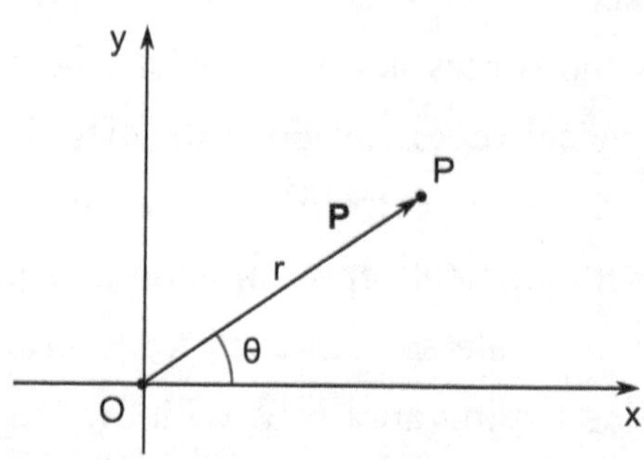

Figure 2.4b. Position vector **P** of point P

The magnitude r of the position vector **P** for the point P is the distance from the origin and the direction is indicated by the angle θ from the x-axis.

The change in position of a point relative to a previous position is described with a vector called displacement vector. In the subsequent figure, the point P represented by position vector **P** is moved to the point P′ represented by position vector **P′**. The displacement vector is $\Delta\mathbf{P} = \mathbf{P}' - \mathbf{P}$.

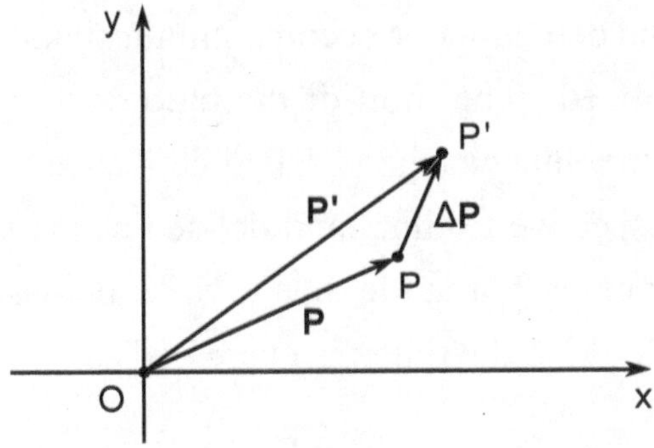

Figure 2.4c. Displacement vector $\Delta\mathbf{P}$

Velocity is the rate of change of position, and the position change is expressed by the position vector change, which is the displacement vector. For the position change, the position vector is a function of time. Assume that at time t, the position vector is $\mathbf{P}$ and at time t', the vector is $\mathbf{P}'$. The average velocity over the time span $\Delta\mathrm{t} = \mathrm{t}' - \mathrm{t}$ is

$$\overline{\mathbf{V}} = \frac{\Delta\mathbf{P}}{\Delta\mathrm{t}}$$

For those who are not familiar with the concept of derivative, the instantaneous velocity described next serves as a good example. In the equation above, if $\Delta\mathrm{t}$ becomes smaller, the average velocity is closer to the velocity at the point P corresponding to the time t. If the $\Delta\mathrm{t}$ is set to a constant value, no matter how small, the average velocity is not the exact velocity at the time t. If the $\Delta\mathrm{t}$ is set to zero, there is no change for the $\Delta\mathrm{t}$ and $\Delta\mathbf{P}$, and the ratio is $\frac{0}{0}$, which is undefined. To solve this dilemma, the concept of limit is introduced, which provides a dynamic and infinitely small change value, which is close to 0 but smaller than any given value. The magic of the limit concept is that as the numerator and the denominator become infinitesimally small, their ratio is determined. The limit of the ratio of the position vector change to the time change is called derivative of the position vector with respective to time, and defined as the instantaneous velocity at the point P or at the time t. Notation wise, the instantaneous velocity **V** at the time t is expressed as

$$\mathbf{V} = \lim_{\Delta t \to 0} \frac{\Delta \mathbf{P}}{\Delta t} = \frac{d\mathbf{P}}{dt}$$

The notation $\frac{d\mathbf{P}}{dt}$ is called derivative of position vector **P** with respective to time t, and represents the instantaneous rate of change of position vector with respective to time.

To see a concrete example of the limit with definite result for the ratio of two infinitesimals, we may consider one-dimensional motion of an object with its position time-dependence function as $x = t^2$, where x is the coordinate for the position and t the time. It is noted that in one dimension, the x coordinate relative to the origin already indicates the direction, and the explicit vector notation is not needed. At time t, the position is $x = t^2$, and at time $t' = t + \Delta t$, the position is $x' = (t + \Delta t)^2$. Thus, $\Delta x = x' - x = 2t\Delta t + (\Delta t)^2 = \Delta t(2t + \Delta t)$. It is important to know that in the limit towards zero, Δt is infinitesimally small and can be practically treated as zero and ignored in comparison with 2t in the addition term $(2t + \Delta t)$, but cannot be done so in the ratio. The derivative or the limit of the ratio is

$$\frac{dx}{dt} = \lim_{\Delta t \to 0} \frac{\Delta x}{\Delta t} = \lim_{\Delta t \to 0} \frac{\Delta t(2t + \Delta t)}{\Delta t} = 2t$$

It can be seen that being infinitesimally small is just relative. In the ratio above, both the numerator and the denominator are infinitesimally small, but when they are compared to each other through the division, their ratio value becomes definite.

The concept of the limit and derivative is of philosophical interest, and is also consistent with the yin-yang implication. For example, on the surface, the function x^2 along with the function curve looks still or static, and there is no never-ending dynamic process involved. But from the yin-yang viewpoint, although the static part (yin) and the dynamic part (yang) are opposite to each other, they can co-exist, interact with and transform into each other. The limit-derivative concept indeed transforms a static function case into a never-ending dynamic process with a surprisingly definite ratio result.

The concept of the limit and derivative also turns out to be very practical and powerful for problem solving and analysis in different branches of mathematics. In general, the derivative represents the instantaneous rate of change of a function at a specific point. We will see later that the instantaneous rate of change of position vector at a point plays an important role in a new discovery related to our main topic.

2.5 Position stretching and position rotation

Next, we will investigate further about displacement vector and associated position movement types. As we may have observed in our daily life, there are two basic types of position movement: position stretching along a straight line and position rotation around a point.

First, let's check the position stretching, which can be realized through multiplication by real numbers. In the case of one-dimensional line, x-axis denotes a real number axis. O is the origin.

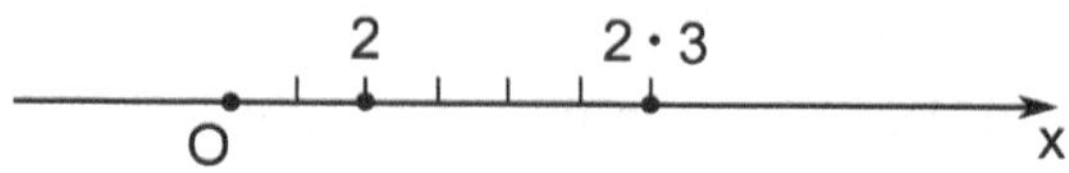

Figure 2.5a. Position stretching by a real number in a line

A point in the axis may be represented with a position number. For example, a point is represented with position number 2. Multiplying 2 by 3 makes the point stretch to the position represented by number 6. It is noted that like a vector, the position number here also implies a direction, which is from the origin to the point. As we can see, the multiplication by a real number only causes the position stretching without direction change.

In the case of two-dimensional plane, the position stretching is demonstrated by the following figure. A point in the plane may be represented with a position vector **P**.

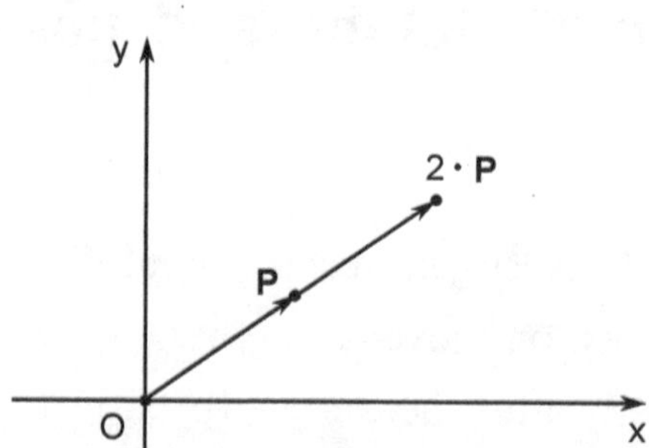

Figure 2.5b. Position vector stretching by a real number in a plane

Multiplying the vector **P** by 2 makes the point stretch to the position represented by position vector 2**P**. Again, as we can see, for vector, the multiplication by a real number also only causes the position stretching without direction change. Here, we may note that the position stretching represents a linear motion.

Now, with regard to the position rotation, we may have noticed that in real life, besides the linear motion, there is another basic motion type that is corresponding to the position rotation. That is, in the physical world, there are two basic types of motion: linear (translational) motion and rotational motion.

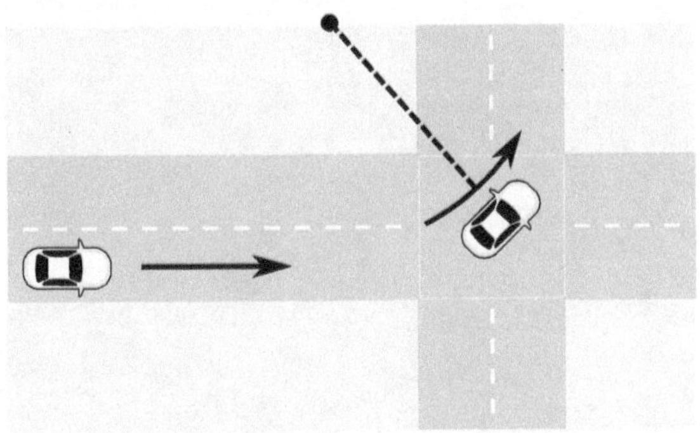

Figure 2.5c. Two basic motion types: linear motion and rotational motion.

As shown in the illustration, for the linear motion, one car moves along a straight line without direction change. The motion may be described with position stretching and multiplication by real numbers. For the rotational motion, another car at the intersection makes a turn or rotation.

Nature tends to be perfect. Then, it is reasonable to think: Rotation is also a basic motion type and there likely exists a new type of number that describes the rotation by multiplication as well. This will be our exploration subject next.

3. Rotation factors as rotational numbers

First, from a high level perspective, one may expect the existence possibility of a new number system for describing rotation. Nature has its laws and properties. Mathematics is invented to discover and describe Nature. Real numbers represent points in a number line, and can describe stretching of a point's position by multiplication. On the other hand, rotation of a point's position is also a fundamental motion type besides the stretching. From the complement and perfection viewpoint, it is natural to make a guess that there likely also exists a number system for the rotation.

Unlike the position stretching, the rotation phenomenon is not confined to a one-dimensional line and requires a two-dimensional plane with the additional dimension. The position of a point in the plane may be represented with a position vector. Real numbers can only make the vector stretch by multiplication. By analogy, rotational numbers are introduced to make the vector rotate by multiplication. Mathematics represents abstractions of properties of the physical world. Rotational numbers may also be called rotation factors in that they all have the same unit length (magnitude) and act as rotating coefficients of directional physical quantities such as position vector. The terms, rotation factor and rotational number are interchangeable. Here, in general, the term rotation factor is mainly used, and the term rotational number is also used under certain context.

3.1 Definition of rotation factors

An angle-dependent rotation factor q of an arbitrary angle δ is defined as a factor for multiplying to a position vector of a point and making the vector rotate counter-clockwise by angle δ relative to the origin point of the vector with the vector's magnitude unchanged. That is

$$\mathbf{P}_\delta = \mathrm{q} \cdot \mathbf{P} = \mathbf{P} \cdot \mathrm{q} \tag{3.1}$$

where (see the figure) position vector $\mathbf{P}$ of point P rotates relative to origin point O by angle δ and reaches position vector $\mathbf{P}_\delta$ of point P_δ.

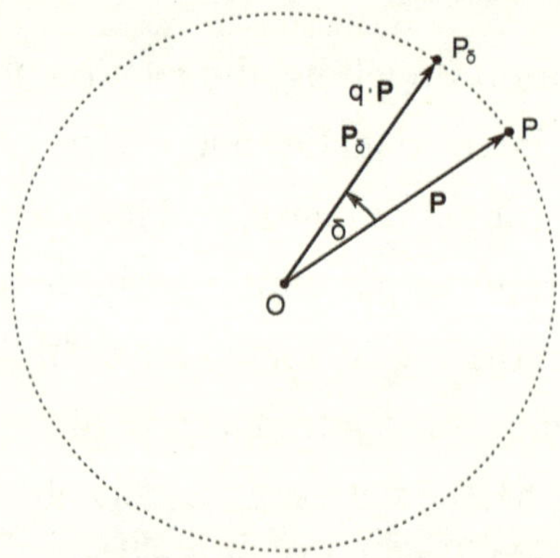

Figure 3.1. Rotation factor q and its target vector **P**

Here, it is convenient to introduce the following terminologies:

Applying a rotation factor to: means multiplying a rotation factor to a position vector. Applying a rotation factor to a point means applying the factor to the point's position vector.

Target of a rotation factor: means the position vector to which a rotation factor is applied.

By the rotation factor definition, the multiplication between a rotation factor and its target vector is commutative. That is, left-multiplication and right-multiplication to a target vector by a rotation factor produce the same rotation result for the target vector.

At this point, the existence of rotation factors such defined is a postulate. If the consequences of the postulate are consistent with existing results, it is then considered true.

3.2 Properties of rotation factors

Denote a rotation factor q of an arbitrary angle δ as $\mathrm{q}(\delta)$. The following basic properties of rotation factors are implied or derived from the rotation factor definition.

1) Being a new type of number with unit magnitude

A rotation factor is not a real number as the former can change a vector's direction by multiplication and the latter cannot. It may be considered as a new type of number. A rotation factor of an arbitrary angle always has a unit magnitude, and will not change the target vector's magnitude.

2) Implying a direction relative to its target vector

It is important to note that a rotation factor of an angle has no direction itself. However, when being applied to a vector, it starts to imply a direction relative to the target vector. The rotation factor will make the target vector rotate to the implied direction.

3) Negative angle representing clockwise rotation and canceling positive angle rotation effect

A rotation factor of angle δ produces a counter-clockwise rotation for a target vector, which may be cancelled by a rotation factor of negative angle $-\delta$ and vice versa. If a rotation factor produces no rotation, it is corresponding to 1. That is

$$\mathrm{q}(\delta) \cdot \mathrm{q}(-\delta) = 1$$

$$\mathrm{q}(-\delta) \cdot \mathrm{q}(\delta) = 1$$

4) Rotation factors being commutative when being applied to a vector

The order of rotation factors for multiplying to a vector will not affect the final result. That is

$$\mathrm{q}(\delta_1) \cdot \mathrm{q}(\delta_2) = \mathrm{q}(\delta_2) \cdot \mathrm{q}(\delta_1)$$

5) Multiplications to a vector by multiple rotation factors being equivalent to one rotation factor with the sum of the individual angles

If the total number of the multiple rotation factors is n. This property is expressed by

$$\mathrm{q}(\delta_1) \cdot \mathrm{q}(\delta_2) \cdot ... \cdot \mathrm{q}(\delta_n) = \mathrm{q}(\delta_1 + \delta_2 + ... + \delta_n)$$

6) Division operation of rotation factor being allowed

The multiplication effect produced by a rotation factor may be inversed through division by the same rotation factor and vice versa. This means

$$\mathrm{q} \cdot \frac{1}{\mathrm{q}} = 1$$

$$\frac{1}{\mathrm{q}} \cdot \mathrm{q} = 1$$

7) Having a period of 2π

After a 2π rotation, a vector's direction remains the same. That is

$$\mathrm{q}(\delta + n \cdot 2\pi) = \mathrm{q}(\delta)$$

where n is an integer.

8) Becoming real numbers 1 and -1 respectively for angles 0 and π

In general, a rotation factor of an arbitrary angle is not a real number. But the two angles 0 and π are the exceptions. The rotation factor of angle 0 produces no rotation for its target vector and becomes the real number 1. And the rotation factor of angle π produces a reverse direction for its target vector by a 180° rotation and becomes the real number -1. That is

$$q(0) = 1$$

$$q(\pi) = -1$$

Furthermore, assume that angle δ times integer n equals the full circle angle 2π ($n \cdot \delta = 2\pi$; $\delta = \frac{2\pi}{n}$) and n is an even number. After n times of multiplication operations to a vector by the angle's rotation factor, the final effect is no rotation as the vector rotates by a full circle angle and comes back to the initial direction. This is equivalent to the final rotation factor being 1. Similarly, after $\frac{n}{2}$ times of the multiplication operations, the final effect is that the vector is rotated by π and the vector's direction is reversed. This is equivalent to the final rotation factor being -1. The two cases are respectively expressed by

$$q(\frac{2\pi}{n})^n = 1$$

$$q(\frac{2\pi}{n})^{n/2} = -1$$

9) Orthogonal rotation factor

Since the two angles 0 and π are exceptional and respectively corresponding to rotation factors being the real numbers 1 and -1, it makes sense to think that the exact middle angle $\frac{\pi}{2}$ between 0 and π is also special.

The rotation factor of the angle $\frac{\pi}{2}$ is here defined as orthogonal rotation factor in that it produces a 90° rotation for the target vector with the resultant direction being perpendicular or orthogonal to the initial direction. Notation-wise, the orthogonal rotation factor is denoted as $\mathrm{q}(\frac{\pi}{2})$, which may be further succinctly denoted by a symbol i. That is

$$i = \mathrm{q}(\frac{\pi}{2}) \tag{3.2}$$

As we will soon see, the orthogonal rotation factor is fundamentally important.

3.3 Fundamental equation of position vector rotation

Since a rotation factor is involved in a position vector's rotation, it is instinctive to explore the rotation motion of the position vector.

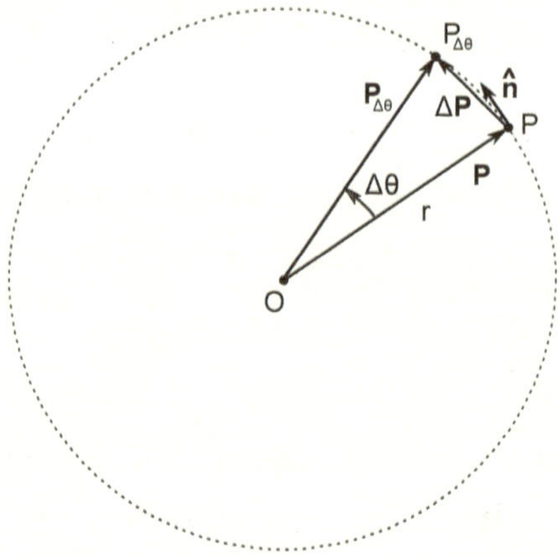

Figure 3.3. Position vector rotation relative to an origin point

In the figure, point P is represented by position vector $\mathbf{P}$ with magnitude r. $\hat{\mathbf{n}}$ is a unit vector perpendicular to the position vector $\mathbf{P}$. Point P rotates around origin point O by angle $\Delta\theta$ with the vector's magnitude r unchanged, and reaches point $P_{\Delta\theta}$ whose position vector is $\mathbf{P}_{\Delta\theta}$. $\mathbf{\Delta P}$ is the vector change, which is the difference between vector $\mathbf{P}_{\Delta\theta}$ and vector $\mathbf{P}$.

Since the vector's magnitude r is unchanged, point P moves along the circumference of the circle during the rotation. Let Δs be the length of the arc corresponding to the angle change $\Delta\theta$, and Δl be the length of the chord of the arc. The Δl is also the magnitude of the vector change $\mathbf{\Delta P}$. Denote $\hat{\mathbf{u}}$ as the unit vector of $\mathbf{\Delta P}$. Then

$$\mathbf{\Delta P} = \|\mathbf{\Delta P}\|\, \hat{\mathbf{u}} = \Delta l \hat{\mathbf{u}}$$

where $\|\mathbf{\Delta P}\|$ is a mathematical notation for the magnitude of the vector $\mathbf{\Delta P}$.

For a circle, the ratio of the arc length to the arc angle span

$(\Delta\theta)$ is always equal to the radius regardless of the value of the angle span. That is, $\frac{\Delta s}{\Delta\theta} = r$ or $\Delta\theta = \frac{\Delta s}{r}$. The ratio of the vector change to the angle change is expressed as

$$\frac{\mathbf{\Delta P}}{\Delta\theta} = (\frac{\Delta \mathrm{l}}{\Delta s})r\hat{\mathbf{u}}$$

As $\Delta\theta \to 0$, $(\frac{\Delta \mathrm{l}}{\Delta s}) \to 1$ and $\hat{\mathbf{u}} \to \hat{\mathbf{n}}$, the limit of the above equation for the derivative with respect to θ leads to

$$\frac{\partial \mathbf{P}}{\partial\theta} = r\hat{\mathbf{n}}$$

where the notation $\frac{\partial \mathbf{P}}{\partial\theta}$ is called partial derivative of position vector $\mathbf{P}$ with respective to θ. Similar to the previously discussed notation $\frac{\mathrm{d}\mathbf{P}}{\mathrm{dt}}$ for the instantaneous rate of change of position vector with respective to time, the partial derivative $\frac{\partial \mathbf{P}}{\partial\theta}$ represents the instantaneous rate of change of position vector with respective to the angle θ. In general, if a function depends on more than one variable, a partial derivative with respective to a particular variable indicates that other variables are treated as constant when the derivation operation is conducted.

It is noted that multiplying the orthogonal rotation factor $\imath$ to the vector $\mathbf{P}$ makes the vector perform an orthogonal rotation with the vector's magnitude r unchanged. In other words, the resultant vector has the magnitude r and the direction $\hat{\mathbf{n}}$. That is

$$\imath\mathbf{P} = r\hat{\mathbf{n}}$$

which along with the previous equation gives

$$\frac{\partial \mathbf{P}}{\partial\theta} = \imath\mathbf{P} \tag{3.3}$$

This serves as a fundamental equation of position vector rotation. It is interesting and striking to note that the orthogonal rotation factor i happens to appear in this fundamental equation.

3.4 Position numbers

As discussed before, for one-dimension, we use the real number system to represent the position of any specific point in a line. By the same logic, for two-dimensions, we will introduce a number system to represent the position of any specific point in a plane.

Referring to the figure below, in a plane, real number r represents point $\mathrm{P_r}$ in the x-axis, and has an implied direction, which is the same as that of the positive x-axis. Therefore, the real number r in the x-axis can be treated as a position vector, and rotated by a rotation factor.

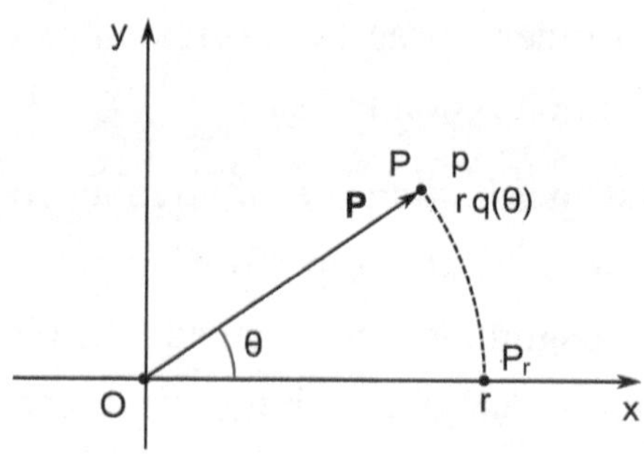

Figure 3.4. Position number

Multiplying the rotation factor $\mathrm{q} = \mathrm{q}(\theta)$ of angle θ to the real number r makes its implied direction rotate by angle θ with magnitude r unchanged. As a result, point $\mathrm{P_r}$ is rotated to point

P, which is represented by an entity p. That is

$$\mathrm{p} = \mathrm{r}\mathrm{q}(\theta) \tag{3.4a}$$

The same point P is represented by position vector **P** as well.

As mentioned before, a rotation factor is a new type of number. Thus, the entity p as the product of a real number and a rotation factor is also a new type of number. Since point P is represented with a position vector, it is natural and lucid to use the counterpart term, position number as the naming for the entity p defined in (3.4a).

Let's compare position numbers and position vectors. As illustrated in the figure, position vector **P** has an explicit direction with an arrow pointing to the direction of the vector. Position number p represents the same point P and also has the same direction (i.e., from the origin point O to the point P). For the position number, the direction is implied without the explicit direction arrow. Both have the same direction and magnitude r. Previously, a rotation factor of an arbitrary angle is applied to a position vector, which has a direction. In fact, the same rotation factor concept can also be applied to a position number. The rotation factor will generate the same rotation for the position number regardless of the target's direction being explicit or implied.

Since a position number and its corresponding position vector have the same direction and magnitude, using the same methodology and Figure 3.3 with position vector notation **P** replaced by position number p, gives the fundamental equation of position number rotation as

$$\frac{\partial \mathrm{p}}{\partial \theta} = i\mathrm{p} \tag{3.4b}$$

We will solve this fundamental equation next.

3.5 Angle dependence formula of rotation factors

To solve the fundamental equation, the dependence of the position number p on the rotation angle θ is needed, and has previously been provided by the position number definition $\mathrm{p} = \mathrm{rq}(\theta)$. With the dependence, the fundamental equation becomes

$$\frac{\partial(\mathrm{rq})}{\partial \theta} = i r \mathrm{q}$$

where $\mathrm{q} = \mathrm{q}(\theta)$.

Since q depends on θ only and r is considered constant with respect to change in θ, it follows that $\frac{\partial(r\mathrm{q})}{\partial\theta} = r\frac{\partial \mathrm{q}}{\partial\theta} = r\frac{\mathrm{dq}}{d\theta}$. Thus, we have

$$\frac{\mathrm{dq}}{d\theta} = i\mathrm{q}$$

From the previous discussion, division is allowed for rotation factor including the orthogonal rotation factor i. Then, the equation becomes

$$\frac{\mathrm{dq}}{d(i\theta)} = \mathrm{q}$$

Let $\phi = i\theta$. The above equation is expressed as

$$\frac{dq}{d\phi} = q$$

Hence, q is the exponential function with its derivative being itself. That is, $q = e^{\phi}$. With $\phi = i\theta$ and $q = q(\theta)$, it follows that

$$q(\theta) = e^{i\theta} \tag{3.5}$$

This result may be called rotation factor formula, which gives the rotation factor dependence on angle. The formula shows the existence of rotation factors covering all angles and has significant implications.

3.6 Developing complex numbers and Euler's formula from first principles

Here, from first principles through the definition of rotation factors and the obtained rotation factor formula, we will proceed to construct a Cartesian coordinate system and develop complex numbers as well as Euler's formula.

Referring to the following figure, in a plane, real numbers a, b, and r represent points in the x-axis.

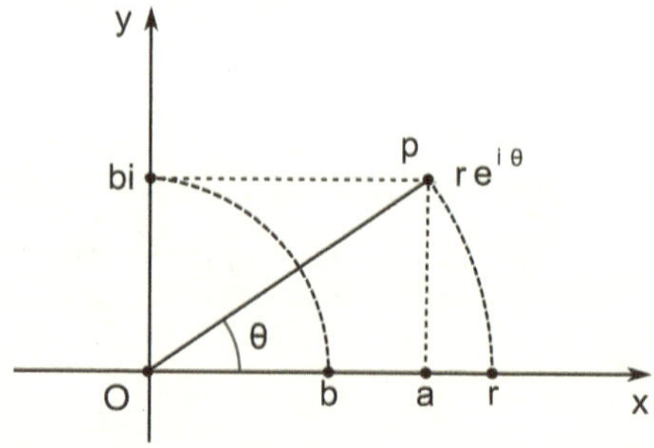

Figure 3.6. Cartesian coordinate system formed from real numbers and rotation factors

It is noted that in general, a point in the x-axis may be considered as a vector with an implied direction that is same as the x-axis unit vector (positive x-axis direction) times the sign of the number representing the point. Similarly, a point in the y-axis has an implied direction that is the same as the y-axis unit vector times the sign of the representing number.

Applying a rotation factor of an arbitrary angle θ to number r makes the point rotate by angle θ to the point represented by position number p or $\mathrm{r}e^{i\theta}$. That is

$$\mathrm{p} = \mathrm{r}e^{i\theta} \tag{3.6a}$$

The line through points a and p is parallel to the y-axis. The distance between the points O and b is the same as the distance between the points a and p. Applying the orthogonal rotation factor i to number b makes the point rotate to the point bi in the y-axis. In the Cartesian coordinate system, the position number p is expressed as

$$\mathrm{p} = a + bi \tag{3.6b}$$

where the terms a and bi are orthogonal to each other, with the former direction pointing to the x-axis unit vector times the sign of a and the latter direction to the y-axis unit vector times the sign of b.

From the definition of sine and cosine, $a = \mathrm{r}\cos(\theta)$ and $b = \mathrm{r}\sin(\theta)$. Thus, (3.6b) becomes $\mathrm{p} = \mathrm{r}\cos(\theta) + i\mathrm{r}\sin(\theta)$, which, together with (3.6a), leads to

$$e^{i\theta} = \cos(\theta) + i\sin(\theta) \tag{3.6c}$$

The result (3.6c) is Euler's formula. The position number p in (3.6b) is equivalent to the definition of a complex number. All complex numbers in the two-dimensional plane form the complex number set $\mathbb{C}$. The orthogonal rotation factor i here is equivalent to the imaginary unit. This shows that we can develop the complex number system including Euler's formula in a new way based on rotation factors without even using and relying on $i = \sqrt{-1}$ or $i^2 = -1$.

4. Orthogonal rotation factor

It is insightful to investigate the orthogonal rotation factor i as it is equivalent to the imaginary unit, which has historically been the base for building the complex number system.

4.1 Properties of orthogonal rotation factor

With reference to the figure, point P relative to origin point O forms its position vector $\mathbf{P}$.

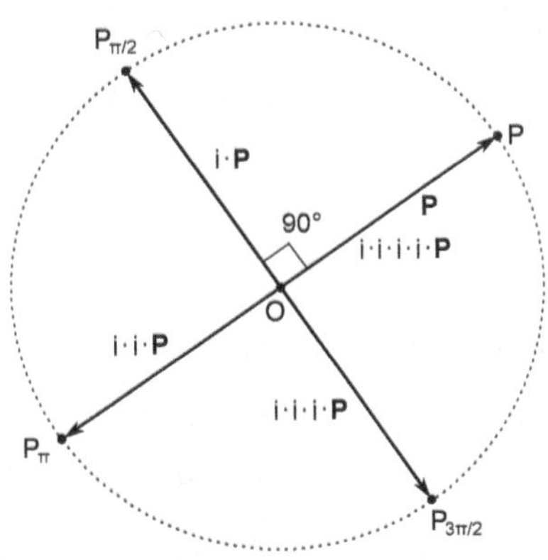

Figure 4.1. Orthogonal rotation factor i being successively applied to position vector

Multiplying the orthogonal rotation factor i to vector $\mathbf{P}$ makes the vector rotate by angle $\pi/2$ or 90° with the resultant vector indicated by $i \cdot \mathbf{P}$ and the resultant point position indicated by point $P_{\pi/2}$.

Further, multiplying i to vector $i{\cdot}\mathbf{P}$ makes the vector rotate by 90° with the resultant vector indicated by $i{\cdot}i{\cdot}\mathbf{P}$ and the resultant point position indicated by point P_{π}. It is noted that the resultant vector has a reversed direction relative to the initial position vector $\mathbf{P}$. That is, $i \cdot i \cdot \mathbf{P} = -\mathbf{P}$ or $i \cdot i = -1$, which is $i^2 = -1$.

And further, multiplying i to vector $i{\cdot}i{\cdot}\mathbf{P}$ makes the vector rotate by 90° with the resultant vector indicated by $i \cdot i \cdot i \cdot \mathbf{P}$ and the resultant point position indicated by point $P_{3\pi/2}$. The resultant vector has a reversed direction relative to the position vector $i \cdot \mathbf{P}$. That is, $i \cdot i \cdot i \cdot \mathbf{P} = -i \cdot \mathbf{P}$ or $i^3 = -i$.

And finally, multiplying i to vector $i \cdot i \cdot i \cdot \mathbf{P}$ makes the vector rotate by 90° with the resultant vector indicated by $i \cdot i \cdot i \cdot i \cdot \mathbf{P}$ and the resultant point position indicated by P. It is noted that the resultant vector is the same as the initial position vector $\mathbf{P}$. That is, $i \cdot i \cdot i \cdot i \cdot \mathbf{P} = \mathbf{P}$ or $i^4 = 1$.

It is important to note the case where $i \cdot i \cdot \mathbf{P}$ = - $\mathbf{P}$ or

$$i^2 = -1 \tag{4.1}$$

This result is an important property of the orthogonal rotation factor. The property also directly shows that the orthogonal rotation factor is not a real number as the square of a real number is always positive. The result has its counterpart for clockwise orthogonal rotation, which is derived next. In the previous Figure 4.1, it is apparent that $\mathrm{q}(-\frac{\pi}{2}) \cdot \mathbf{P} = -\mathrm{q}(\frac{\pi}{2}) \cdot \mathbf{P}$. This means that vector $\mathbf{P}$ rotated clockwise by $\frac{\pi}{2}$ has the opposite direction to vector $\mathbf{P}$ rotated counter-clockwise by $\frac{\pi}{2}$. Therefore, $\mathrm{q}(-\frac{\pi}{2}) = -\mathrm{q}(\frac{\pi}{2})$ or $\mathrm{q}(-\frac{\pi}{2}) = -i$. This states that $-i$ is equivalent to the rotation factor that makes a vector rotate clockwise by $\frac{\pi}{2}$. Multiplying $(-i)$ twice to $\mathbf{P}$ makes the vector rotate by 180° clockwise with the resultant direction being reversed. That is, $(-i) \cdot (-i) \cdot \mathbf{P} = -\mathbf{P}$ or $(-i) \cdot (-i) = -1$, which leads to

$$(-i)^2 = -1$$

4.2 Acting as rotation generator and unit vector

We have shown that orthogonal rotation factor is equivalent to the imaginary unit i, which was originally defined as $\sqrt{-1}$ and coined in the 17th century by René Descartes as a derogatory term and regarded as fictitious or useless. An imaginary number is defined as a real number b multiplied by the imaginary unit i in the form of bi, and a complex number is defined as the addition of an imaginary number to a real number a in the form of $a + bi$. Historically, the imaginary unit has caused an agonizing and slow adoption of complex numbers by mathematicians due to the fictitious nature. Even today, as it is still based on the same very abstract concept of $\sqrt{-1}$ or $i^2 = -1$, the conundrum persists, hindering and discouraging many learners.

Now, with the introduction of rotation factors, the imaginary unit can be directly connected to the physical world through the equivalence of the orthogonal rotation factor, which acts as both an orthogonal rotation generator for rotating vectors and a unit vector pointing to a direction orthogonal to a reference direction.

In the subsequent figure, a person points his arm horizontally forward, and then rotates and points the arm vertically upward, with the orthogonal rotation factor i representing the unit vector for the resultant vertical direction. Here, the initial horizontal direction with the associated vector length (from shoulder joint to pointer fingertip) is assigned as the unit vector 1.

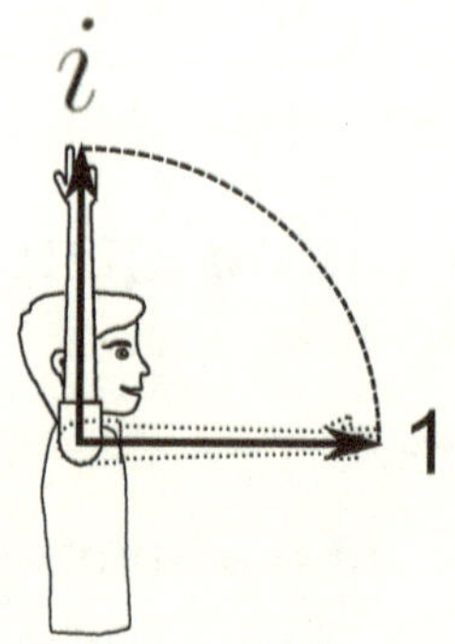

Figure 4.2a. Exhibition of orthogonal rotation factor i by arm rotation

As an illustration for orthogonal rotation factor acting as both a 90° rotation generator and a unit vector, real number 1 implies a direction relative to the origin (shoulder joint in the figure) and may be considered as a unit vector. When being multiplied to 1, the orthogonal rotation factor i acts as a rotation generator and makes the unit vector 1 rotate counter-clockwise by 90°. The resultant direction is i. The process is described by the equation $1 \cdot i = i$. The i on the left-side of the equation acts as the rotation generator, and the i on the right-side represents the resultant direction and acts as the unit vector.

In this figure for an x-y Cartesian coordinate system,

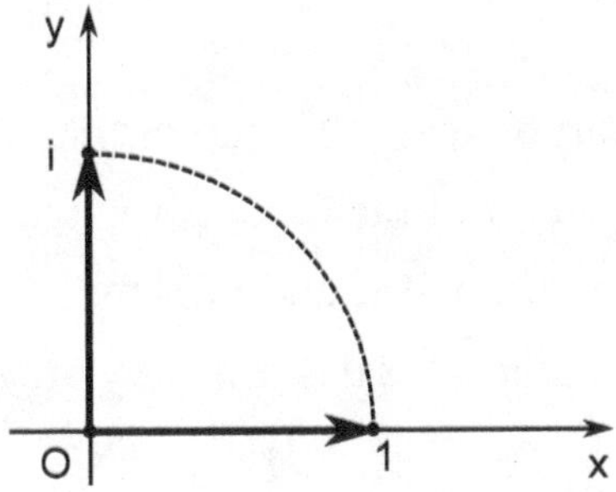

Figure 4.2b. Number 1 as x-axis unit vector and i as y-axis unit vector

the unit vector 1 for the x-axis direction is rotated by orthogonal rotation factor i and becomes the unit vector i for the y-axis direction. In practice, the directions of the unit vectors are implied. The unit vector 1 is replaced with real number 1, and the unit vector i is replaced with position number i. But, in our mind, we may still consider real number 1 as x-axis unit vector and i as y-axis unit vector.

4.3 Participating in algebraic operations

Besides acting as a unit vector, the orthogonal rotation factor i also happens to be able to participate in algebraic operations. For the case of multiplication operation, i may act as a rotation generator to rotate vectors or position numbers; or may act as a vector to be rotated by other rotation factors or position numbers. To specify the rotation target, we may define that the former term in the multiplication expression is rotated by the latter. For example, $\mathrm{p} \cdot i$ means that position number p is rotated by i. It is noted that we may also define the other way around. It will not change the result.

As an illustration for geometric representation of the multiplication $i \cdot i$, the unit vector of the first i is rotated by the orthogonal rotation factor of the second i, and becomes the unit vector -1. Mathematically, the process is described by the formula $i{\cdot}i = -1$.

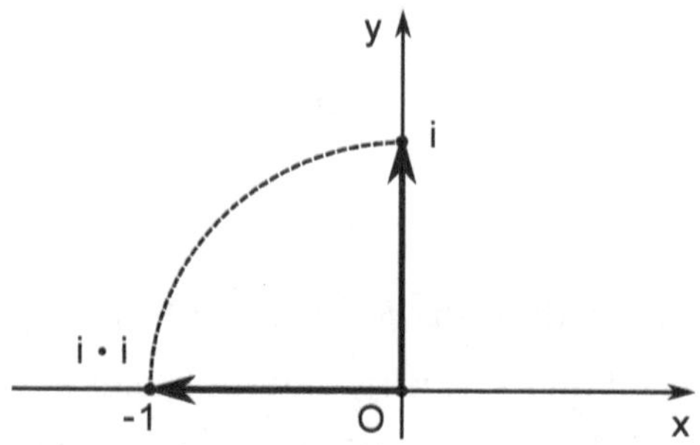

Figure 4.3a. Unit vector i rotated by orthogonal rotation factor i

Further, for the multiplication $(1 + i) \cdot i$ as shown in the following figure,

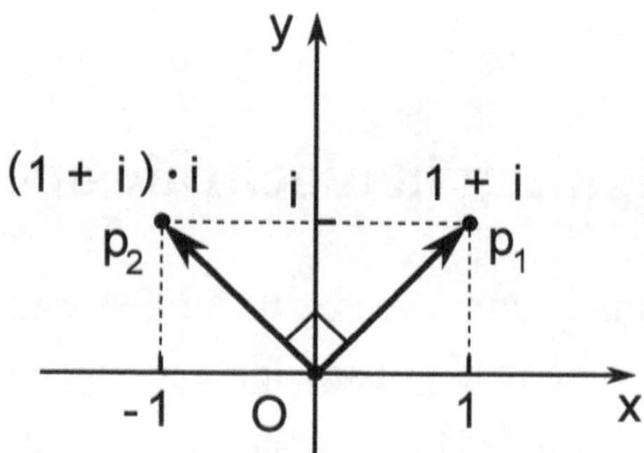

Figure 4.3b. Position number $1 + i$ rotated by orthogonal rotation factor i

position numbers p_1 and p_2 are denoted with vectors to explicitly indicate their directions. The position number $\mathrm{p}_1 = 1 + i$ is rotated by the orthogonal rotation factor i. After the 90° rotation, the point represented by p_1 reaches the point represented by the position number p_2, which has an x-component of -1 and a y-component of i. The process is described by the equation $(1+i) \cdot i = -1 + i$.

As a multiplication-commutative example, the subsequent figure illustrates the multiplication $i \cdot (1 + i)$.

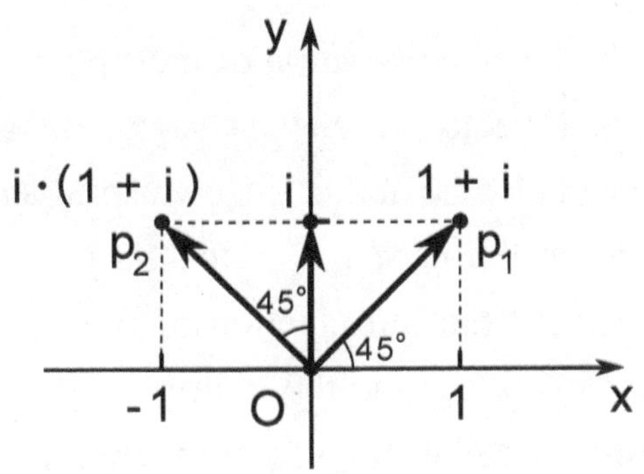

Figure 4.3c. Unit vector i rotated by position number $1 + i$

The position number $p_1 = 1 + i$ has a magnitude of $\sqrt{2}$ and an angle of 45° from the x-axis, which is corresponding to a rotation factor of 45° or $\pi/4$. Therefore, the multiplication $i \cdot (1 + i)$ means that the unit vector i is rotated counter-clockwise by angle 45° and stretched in length by $\sqrt{2}$. That is, the multiplication makes the unit vector i becomes the vector represented by the position number p_2, which also has an x-component of -1 and a y-component of i. The result is described by the equation $i \cdot (1 + i) = -1 + i$.

We may obtain the same result by direct algebraic manipulation: $i \cdot (1 + i) = i + i \cdot i = -1 + i$.

4.4 Division of vectors

As will be seen later, the division of one vector by another may be used to define the constant for the imaginary unit i, i.e., the orthogonal rotation factor. Before using the vector division concept, it helps to provide a proper definition for division of vectors.

Division is the inverse operation of multiplication. If there is multiplication for a vector, a relevant vector division as the inverse operation can also be defined. How useful a defined vector division depends on the need for us to find out the division of two vectors. Without the introduction of rotation factors, only real numbers are involved, and the multiplication of a vector by a real number only stretches the vector in length. The division of a vector by another vector results in a real number, which is the ratio of the lengths (magnitudes) of the two vectors. Now with the introduction of rotation factors, besides the length stretching, a vector may be rotated through multiplication by a rotation factor. The division of vectors becomes more sophisticated, and it is then more useful to provide the definition for the vector division.

Before obtaining the inverse for the division, we may check the multiplication process for two-dimensional vectors.

Referring to the figure,

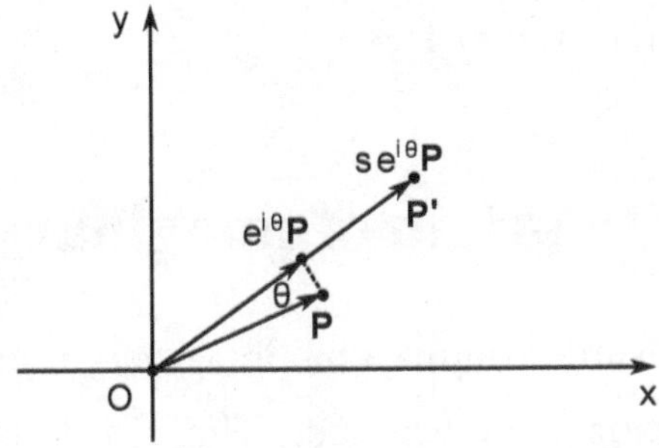

Figure 4.4. Vector $\mathbf{P}$ rotated by a rotation factor $e^{i\theta}$ and stretched by a scaling factor s with the resultant vector denoted by $\mathbf{P}'$.

in a plane, vector $\mathbf{P}$ represents a point. Multiplying a rotation factor of angle θ makes vector $\mathbf{P}$ rotate to the point represented by vector $e^{i\theta}\mathbf{P}$. Multiplying a real number scaling factor s makes vector $e^{i\theta}\mathbf{P}$ stretch to the point represented by vector $se^{i\theta}\mathbf{P}$, which is the resultant vector $\mathbf{P}'$. The multiplication process is represented by the following formula

$$\mathbf{P}' = s \cdot e^{i\theta} \cdot \mathbf{P}$$

By the inverse of the multiplication process, we may define the division of vectors as follows

$$\frac{\mathbf{P}'}{\mathbf{P}} = s \cdot e^{i\theta} \tag{4.4}$$

This formula provides the vector division definition for two dimensions. The vector division result gives the information about the numerator vector $\mathbf{P}'$ relative to the denominator vector $\mathbf{P}$. Specifically, the division result is a position number with the magnitude s and the rotation factor $e^{i\theta}$. The magnitude s as a real number indicates the length of vector $\mathbf{P}'$ relative to that of vector $\mathbf{P}$. The rotation factor $e^{i\theta}$ indicates the rotation angle of vector $\mathbf{P}'$ relative to vector $\mathbf{P}$.

4.5 Causality origin of imaginary unit i

Next, the causality origin of the imaginary unit or the orthogonal rotation factor i is discussed. The i is defined in (3.2) as $\mathrm{q}(\frac{\pi}{2})$, the rotation factor of angle $\frac{\pi}{2}$. Up to now, the i is still a symbol as $\mathrm{q}(\frac{\pi}{2})$. One may still ask what is $\mathrm{q}(\frac{\pi}{2})$? To answer this question, it is noticed that $\mathrm{q}(\frac{\pi}{2})$ is a constant that is denoted by i. And π is also a constant. From the causality viewpoint, investigating the cause of the π may provide a hint for the cause of the i.

In this figure, point O is the origin of position vector $\mathbf{P}$, whose instant rotation is represented by $\frac{\partial \mathbf{P}}{\partial \theta}$.

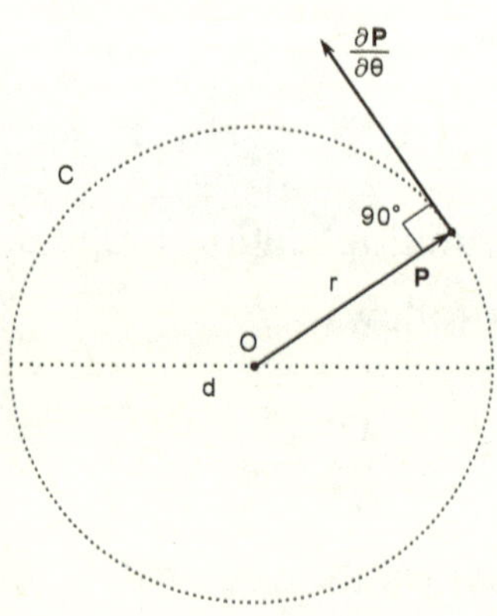

Figure 4.5. Illustration of π as ratio of circle's circumference C to diameter d and i as ratio of vector $\frac{\partial \mathbf{P}}{\partial \theta}$ to vector $\mathbf{P}$.

The term $\frac{\partial \mathbf{P}}{\partial \theta}$ is also a vector, whose direction is perpendicular to that of the vector $\mathbf{P}$. The magnitude of the vector $\frac{\partial \mathbf{P}}{\partial \theta}$ is the same as that of the vector $\mathbf{P}$. The associated circle for the rotation of the position vector $\mathbf{P}$ is also illustrated in the figure. The circle is centered at the origin O and has the circumference C and diameter d as well as the radius r, which is also the magnitude of both $\mathbf{P}$ and $\frac{\partial \mathbf{P}}{\partial \theta}$.

Since the ratio of a circle's circumference C to its diameter d is a constant, this leads to the definition of π. That is

$$\pi = \frac{\mathrm{C}}{\mathrm{d}}$$

where C and d are variables, but their ratio π is a constant.

By the same token, from the fundamental equation of position vector rotation, the constant i may be defined as

$$i = \frac{\frac{\partial \mathbf{P}}{\partial \theta}}{\mathbf{P}} \tag{4.5}$$

where the derivative of position vector with respective to rotation angle $\frac{\partial \mathbf{P}}{\partial \theta}$ and the position vector $\mathbf{P}$ are equivalent to variables, but their ratio is a constant. It is the existence of the ratio being constant that causes the existence of the i and the existence of the rotation factor formula $\mathrm{q}(\theta) = e^{i\theta}$, which indicates the existence of the rotation factor set, as will be discussed later. It is important to point out that the term $\frac{\partial \mathbf{P}}{\partial \theta}$ and the term $\mathbf{P}$ have their respective directions, which are perpendicular to each other. As a result, their ratio i is a special non-real number constant and represents an orthogonal rotation.

From the perspective of dimensions, there is also a similarity between π and i. In $\pi = \frac{\mathrm{C}}{\mathrm{d}}$ for π, the diameter d is a one-dimension line, and the circumference C is a two-dimension circle. The constant π is related to the ratio of a variable in two-dimensions to a variable in one-dimension. And in (4.5) for i, the $\mathbf{P}$ itself is in a one-dimension line. The rotation motion is represented by the derivative $\frac{\partial \mathbf{P}}{\partial \theta}$, which is in two-dimensions relative to the $\mathbf{P}$. Thus, the constant i is also related to the ratio of a variable in two-dimensions to a variable in one-dimension.

The i such defined in (4.5) has been shown to be able to derive the Euler's formula, become the equivalent of the imaginary unit, and have the properties of $i^2 = -1$ and $(-i)^2 = -1$, which can serve as the solutions to the equation $z^2 + 1 = 0$.

Both π and i are important mathematical constants. It is amazing to see that both share the similarities and connections through circle and rotation. Here we may feel in awe about the inherent elegance and magic of Nature exhibited through mathematics.

5. Rotation factor set

In mathematics, a field is a set on which the four basic operations of addition, subtraction, multiplication, and division are defined and behave as the corresponding operations on real numbers. The best known fields are the field of rational numbers and the field of real numbers. The set of real numbers is denoted $\mathbb{R}$.

In comparison, rotation factors are a new kind of number. And all rotation factors of different angles form a set that also conforms to being a field. Next, we will discuss the existence of such a number set as well as important characteristics of the set.

5.1 Existence of rotation factor set

The discovery of the rotation factor formula $\mathrm{q}(\theta) = e^{i\theta}$ reveals the existence of rotation factor set. It should be noted that in the derivation of the rotation factor formula, the rotation angle θ is relative to the x-axis. But, in reality, the angle of a rotation factor in the definition $\mathbf{P}_\delta = \mathrm{q} \cdot \mathbf{P}$ and in the formula can be relative to the target's direction and independent of coordinate systems.

The rotation factors can be used to construct coordinate systems. For a polar coordinate system, from $\mathrm{p} = \mathrm{rq}(\theta)$ and $\mathrm{q}(\theta) = e^{i\theta}$, the position number p is expressed as

$$\mathrm{p} = re^{i\theta}$$

This formula means that multiplying a rotation factor of angle θ to a real number r in an x-axis makes the point of the number rotate to the point represented by the position number p. In other words, the position number is formed by multiplying a real number with a rotation factor.

All rotation factors of different angles represented by the formula $\mathrm{q}(\theta) = e^{i\theta}$ form a field on which the four basic operations can be performed. The set of rotation factors $\mathbb{E}$ is then given as

$$\mathbb{E} = \{e^{i\theta} \mid \theta \in \mathbb{R}, -\pi < \theta \leq \pi\}$$

Here, we don't have to be intimidated by the mathematical notation, which just says that the set $\mathbb{E}$ is formed by all members of $e^{i\theta}$ with the angle θ belonging to real numbers and being in the range of $-\pi < \theta \leq \pi$.

Alternatively, we may define the set $\mathbb{E}$ as

$$\mathbb{E} = \{e^{i\theta} \mid \theta \in \mathbb{R}, 0 \leq \theta < 2\pi\}$$

where the angle θ is in the range of $0 \leq \theta < 2\pi$.

Both definitions above for the $\mathbb{E}$ cover all angles in 360 degrees. The former definition is often used for representing angle changes in a coordinate system, while the latter is more convenient for describing periodic rotations with a period of 2π. It is noted that although important and special, the orthogonal rotation factor i is just one member in the set $\mathbb{E}$.

Let $\mathbb{R}_{\geq 0}$ denote the set of zero and positive real numbers, which is a subset of $\mathbb{R}$ and forms a positive x-axis. As illustrated in the following figure, a plane is completely formed by a 360° or 2π rotation scan of the positive x-axis.

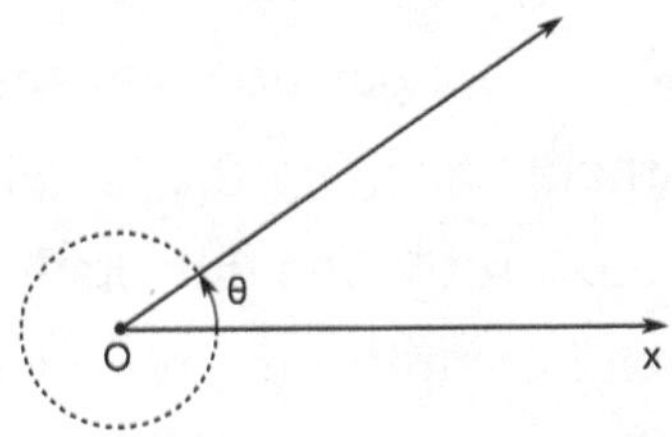

Figure 5.1a. A plane formed by a 2π rotation scan of the positive x-axis

In other words, a plane can be formed by the set $\mathbb{R}_{\geq 0}$ and the set $\mathbb{E}$.

Let $\mathbb{P}$ denote the set of all position numbers. $\mathbb{P}$ represents all points in a plane. A member in $\mathbb{P}$ may be constructed by a member in $\mathbb{R}_{\geq 0}$ and a member in $\mathbb{E}$. This figure illustrates their relationship:

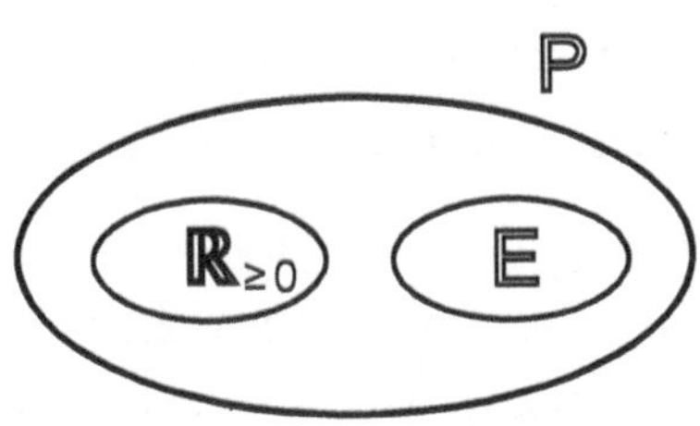

Figure 5.1b. Relationship of non-negative real number set $\mathbb{R}_{\geq 0}$, rotation factor set $\mathbb{E}$, and position number set $\mathbb{P}$

The position stretching and position rotation are two basic motion types in the physical world. From the perspective of describing motion of points in a plane, numbers may be classified into two types: real numbers for position stretching, and rotation factors for position rotation. Multiplying a real number to a position number or position vector stretches the target's magnitude without changing the target's direction. On the other hand, multiplying a rotation factor to a position number or position vector rotates the target without changing the target's magnitude. The existences of real number set $\mathbb{R}$ and the rotation factor set $\mathbb{E}$ complement each other and complete the representation for the two basic motion types. It is believed that if advanced aliens exist, they also have the similar concept about the existence and characteristics of the rotation factors although their mathematics may take radically different forms than ours.

5.2 2D Arithmetic

Discovering the existence of the rotation factor set and understanding its properties and implications give rise to 2D (two dimensional) arithmetic.

The position number set $\mathbb{P}$ represents all points in a 2D plane, while the real number set $\mathbb{R}$ represents all points in a 1D line. The rotation factor set $\mathbb{E}$ covers all angles for rotation with the orthogonal rotation factor i being one of its members for the angle $\frac{\pi}{2}$. Each rotation factor in the set $\mathbb{E}$ has a magic property of rotating a position number or position vector by its angle amount through multiplication. The 2D plane is an abstract and linkage to the physical world. The existence of the rotation factor set is a great gift from Nature and deserves to be cherished for broad learning and utilization.

One of the main benefits of 2D arithmetic is to allow elementary students to use the familiar position and rotation concepts of the physical world and solve immediate mathematics problems from a higher dimension perspective than 1D without the deeper knowledge behind the rotation factors.

Next, how 2D arithmetic may be readily constructed for such a purpose is illustrated. To hide the complexity to a beginner, denote a rotation factor of angle θ as

$$[\theta] = e^{i\theta} \tag{5.2}$$

The square brackets [] appear to be a good selection in terms of being less conflicting with existing conventions, common and direct for typing from a standard computer keyboard. Rotation factors may also be called rotational numbers in direct comparison with real numbers.

Multiplication of a number by a real number stretches the position of the multiplicand, the number being multiplied. By the same token, rotational numbers may be introduced and multiplication of a number by a rotational number rotates the position of the multiplicand. This figure provides an illustration for position rotation in contrast to position stretching:

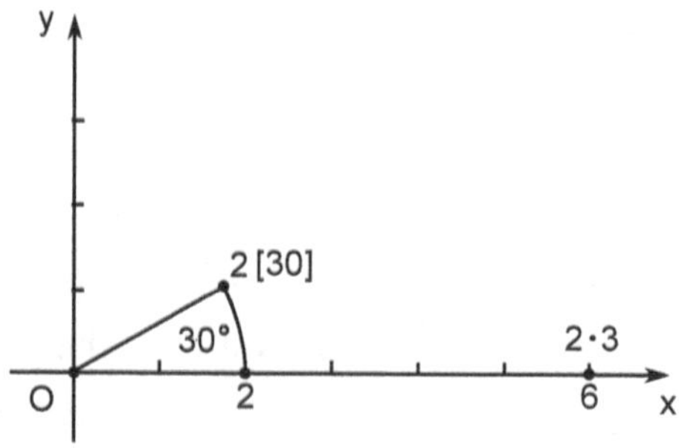

Figure 5.2a. Real number for position stretching versus rotational number for position rotation

Number 2 multiplied by real number 3 makes its position stretch to the position represented by number 6. In contrast, the number 2 multiplied by rotational number [30] makes its position rotate to the position represented by 2[30]. It is noted that in this illustration and others, the degree unit is used for an angle inside the square brackets.

The following figure presents an example of multiplication in 2D arithmetic:

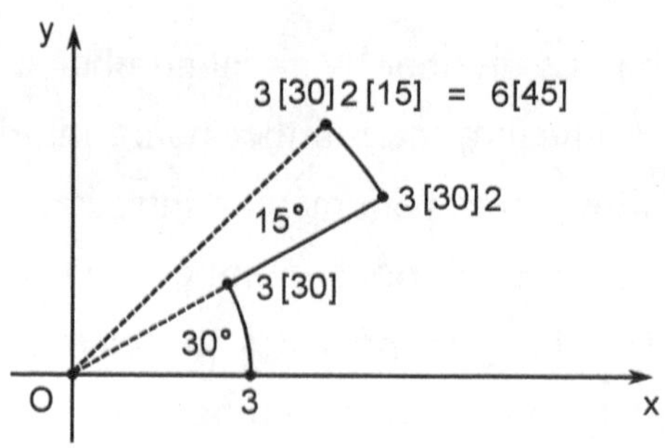

Figure 5.2b. Illustration of multiplication in 2D arithmetic

Multiplying rotational number [30] to the number 3 makes the point rotate by 30 ° relative to the number's direction and reach 3[30]; multiplying the position number 3[30] by 2 makes the point stretch to 3[30]2; and multiplying rotational number [15] to 3[30]2 makes the point rotate to 3[30]2[15], which is equal to 6[45]. It is noted that [30][15] = [30 + 15] = [45], due to the fact that multiple rotations result in a total rotation angle as the sum of the individual rotation angles (which also happens to be magically consistent with the exponential function rule that multiplying multiple exponential functions results in the individual exponents being added). A mobile phone app calculator may be developed and used to calculate rotational numbers such as [45].

It should be mentioned that for rotational numbers, multiplication also represents division operation, which produces a rotation in the opposite direction compared to that of multiplication. For example, [30]/[15] = [30][-15] = [30 - 15] = [15].

Next, we discuss the addition operation in 2D arithmetic where rotational number [90] plays an important role. It should be pointed out that the 90 ° rotational number [90] is the orthogonal rotation factor, and may be denoted by the symbol i:

$$i = [90]$$

The following figure provides an illustration of addition in 2D arithmetic:

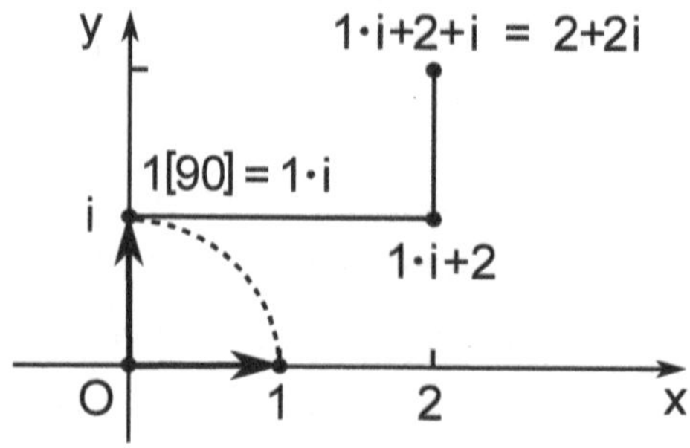

Figure 5.2c. Illustration of addition in 2D arithmetic

The number 1 is considered as the unit vector in the x-direction. Multiplying rotational number [90] to the number 1 makes the point rotate by 90 degrees and reach 1[90], which is equal to $1{\cdot}i$. The i represents the unit vector in the y-direction. Adding the position number $1{\cdot}i$ by 2 makes the point move horizontally in the x-direction to the position $1{\cdot}i + 2$. And adding the position number $1{\cdot}i + 2$ by i makes the point move vertically in the y-direction to the position $1{\cdot}i + 2 + i$, which is equal to $2 + 2i$.

Desirably, we should have embraced and popularized rotational numbers much earlier in that the set of rotational numbers $\mathbb{E}$ already exists in Nature as a treasure to us for easy direct use. The contemporary situation of ubiquitous touch-screen usage by mobile phones indicates that selecting a point on a surface can actually be more primitive and intuitive than selecting a point on a line. This also supports the idea of teaching 2D arithmetic earlier to elementary students.

5.3 Characteristics of rotation factors

The rotation factor dependence on angle is given by the rotation factor formula $\mathrm{q}(\theta) = e^{i\theta}$. That is, a rotation factor of arbitrary angle θ is $e^{i\theta}$. For convenience, the term $e^{i\theta}$ is called the rotation factor function or the rotation factor in short. Rotation factors represented by the rotation factor function have some important characteristics that are identified next.

1) Natural exponential function

The rotation factor $e^{i\theta}$ is the natural or base-e exponential function. Therefore, the derivative of the rotation factor returns the rotation factor itself. This derivative characteristic of the rotation factor has broad applications in mathematics and other sciences.

2) Multiplication of rotation factors equivalent to sum of angles

Multiplications to a vector by multiple rotation factors is equivalent to one rotation factor with the sum of the individual angles. For example,

$$\mathrm{q}(\theta_1)\mathrm{q}(\theta_2) = \mathrm{q}(\theta_1 + \theta_2)$$

With the rotation factor function, the above is expressed as

$$e^{i\theta_1} e^{i\theta_2} = e^{i(\theta_1+\theta_2)}$$

This characteristic converts a multiplication operation into an addition operation, and has many practical applications, particularly in electrical engineering where the angle change in the resultant exponential function is called phase shift.

3) Magnitude of the rotation factor being 1

The rotation factor will not cause the stretching of the target vector. Therefore, the magnitude of the rotation factor $e^{i\theta}$ is always 1.

4) Magnitude of the rotation factor change rate being 1

The rotation factor $\mathrm{q} = e^{i\theta}$ obeys the following equation:

$$\frac{d\mathrm{q}}{d\theta} = i\mathrm{q}$$

The magnitude of the above equation is:

$$\left|\frac{d\mathrm{q}}{d\theta}\right| = |\mathrm{q}| = 1$$

This means that the magnitude of $\frac{d\mathrm{q}}{d\theta}$, which is the rotation factor instant change rate with respect to rotation angle, is always equal to the magnitude of q itself, which is a constant 1 regardless of the rotation angle value. And further the implied direction of the instant change rate $\frac{d\mathrm{q}}{d\theta}$ is always perpendicular to the implied direction of q.

5) The rotation factor requiring a target for rotation

The rotation factor $e^{i\theta}$ itself will not rotate. It needs a target vector or number that has or implies a direction for completing the rotation. When the rotation factor $e^{i\theta}$ is applied to unit number 1 (considered as unit vector), it causes the unit vector to rotate by angle θ from the x-axis without the vector stretching, and forms a unit circle for different rotation angles.

5.4 Unit circle of rotation factors

All rotation factors form the rotation factor set $\mathbb{E}$, which may be represented by a unit circle. Next, we will discuss the unit circle and obtain Euler's formula from the circle. For comparison, we will further present the first method that Euler used to derive the formula.

The unit circle representation is shown in the following figure. Multiplying rotation factor of angle θ makes the point of number 1 rotate to the point p represented by position number $e^{i\theta}$ with the magnitude of the number unchanged and remaining unit 1. Changing the value of the angle θ makes the corresponding point p move along the unit circle.

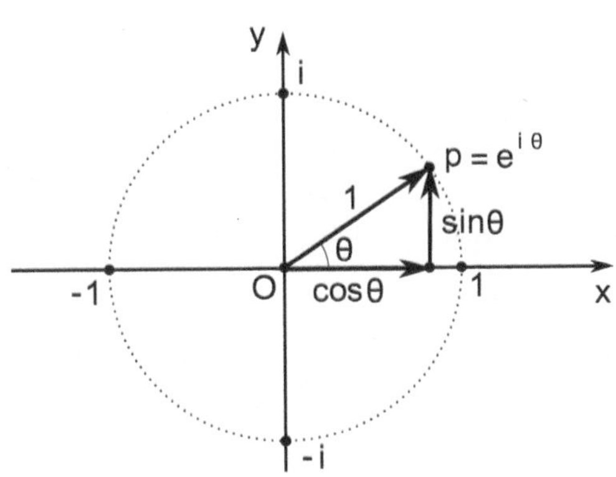

Figure 5.4. Unit circle representation for the rotation factor set

The point p has an implied direction and may be denoted by a vector with arrow. The position number p $= e^{i\theta}$ has its x-component and y-component. The direction of the x-component is the direction of number 1 or that of the positive x-axis, while the direction of the y-component is the direction of i or that of the positive y-axis. By the definition of trigonometric functions, the x-component is $cos(\theta)$, and the y-component is $isin(\theta)$. That is, expressing the position number $e^{i\theta}$ by its components, we have

$$e^{i\theta} = sin(\theta) + icos(\theta)$$

This is the well-known Euler's formula and has been generally considered "the most important formula of mathematics."

From the unit circle and x-y components above, we can directly derive Euler's formula and comprehend its geometric meaning. For comparison, the historical derivation is presented next.

Euler's formula was first derived by Euler using power series, i.e., Taylor series for the exponential function, along with basic facts about the powers of i:

$i^0 = 1$, $i^1 = i$, $i^2 = -1$, $i^3 = -i$, $i^4 = 1$, ...

In mathematics, the Taylor series of a function is an infinite sum of terms that are expressed in terms of the function's derivatives at a single point. If a function $f(x)$ is equal to the sum of its Taylor series for all points x in the complex plane, it is called entire. The exponential function e^x, and the trigonometric functions sine and cosine, are examples of entire functions.

The exponential function can be expressed in the power series as

$$e^x = \sum_{n=0}^{\infty} \frac{x^n}{n!} = 1 + x + \frac{x^2}{2!} + \frac{x^3}{3!} + ...$$

where $n!$ stands for the the factorial of n, with $3! = 3 \cdot 2 \cdot 1$ for example. This power series is also valid for x values that are complex numbers.

Similarly, the sine and cosine functions can be expressed as

$$sin(x) = x - \frac{x^3}{3!} + \frac{x^5}{5!} - \frac{x^7}{7!} + ...$$

$$cos(x) = 1 - \frac{x^2}{2!} + \frac{x^4}{4!} - \frac{x^6}{6!} + ...$$

With the exponential function power series above, for complex values ix with x being real, we have

$$\begin{aligned} e^{ix} &= 1 + ix + \frac{(ix)^2}{2!} + \frac{(ix)^3}{3!} + \frac{(ix)^4}{4!} + \frac{(ix)^5}{5!} + \frac{(ix)^6}{6!} + ... \\ &= \left(1 - \frac{x^2}{2!} + \frac{x^4}{4!} - \frac{x^6}{6!} + ...\right) + i\left(x - \frac{x^3}{3!} + \frac{x^5}{5!} - ...\right) \\ &= cos(x) + isin(x) \end{aligned}$$

which is Euler's formula.

It is amazing to see that the ultra important Euler's formula can be derived from seemly very different methods. This indeed also illustrates the magic connection and consistency of mathematics throughout its various branches.

5.5 Basic mathematical constants in rotation factors

Euler's number e and imaginary unit i are two important basic mathematical constants. The absence of an explanation for the emergence of e in association with i may leave many learners dissatisfied.

The constant e can be expressed in many ways. Jacob Bernoulli introduced it in 1683, for solving the problem of continuous compounding of interest, with its definition as the following limit:

$$e = \lim_{n \to \infty} \left(1 + \frac{1}{n}\right)^n$$

where n indicates how many times per year the interest is compounded (e.g. $n = 12$ for monthly compound interest).

It is also the sum of the infinite series:

$$e = \sum_{n=0}^{\infty} \frac{1}{n!} = 1 + \frac{1}{1} + \frac{1}{1 \cdot 2} + \frac{1}{1 \cdot 2 \cdot 3} + \ldots$$

Moreover, it is the value at 1 of the natural exponential function e^x: e^1. And there are various other expressions.

Interestingly, in contrast to i, which historically first arose from being a solution to a polynomial equation, e is transcendental, meaning that it is not a solution of any non-zero polynomial equation with rational coefficients, according to the Lindemann–Weierstrass theorem. Although e and i are radically different in categories and characteristics, both appear in $e^{i\theta}$. At first glance, this seems surprising and perplexing. But this is the natural consequence of the fundamental equation of position vector rotation, $\frac{\partial \mathbf{P}}{\partial \theta} = i\mathbf{P}$.

The orthogonal rotation factor i, as the equivalent of imaginary unit, appears because the fundamental equation dictates that the direction of the derivative of position vector with respect to rotation angle be orthogonal to the position vector itself. Further, the constant e appears because the derivative of position vector is proportional to the position vector itself, which conforms to the derivative property of the natural exponential function. Specifically, position number p is equivalent to position vector as they have the same direction and magnitude. With rotation factor q rotating number r to reach p, we have $\mathrm{p} = r\mathrm{q}$ and $\frac{d\mathrm{q}}{d\phi} = \mathrm{q}$ where $\phi = i\theta$. This leads to $\mathrm{q} = e^{i\theta}$. Even furthermore, the other important basic constant π also appears because the angle θ is present in the rotation factor $e^{i\theta}$ and a specific angle is often expressed as π or a fraction of π.

As a special case, from the previous discussion for properties of rotation factors, we have $\mathrm{q}(\pi) = -1$. That is, the q value of angle π is known and the rotation factor formula $\mathrm{q}(\theta) = e^{i\theta}$ for the angle gives $\mathrm{q}(\pi) = e^{i\pi}$. Then it follows that $e^{i\pi} = -1$ or

$$e^{i\pi} + 1 = 0$$

This is Euler's identity, which is considered to be an exemplar of mathematical beauty in that it exhibits a profound connection between the five fundamental constants in mathematics: π, i, e, 0 and 1.

The identity is geometrically illustrated by the figure:

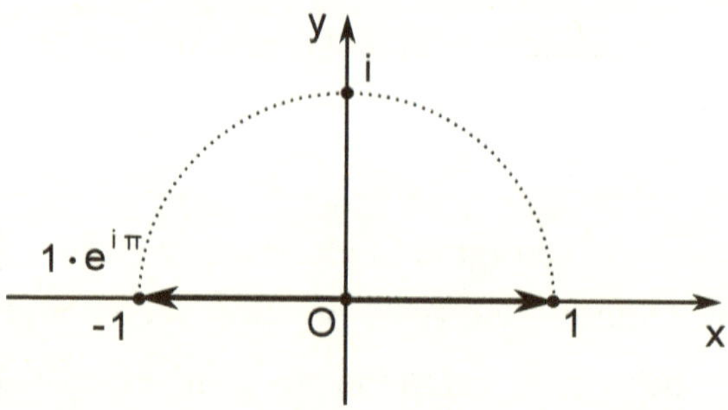

Figure 5.5. Number 1 multiplied by $e^{i\pi}$, the rotation factor of angle π, rotates to the point -1.

The meaning behind this wonderful identity is that multiplying the rotation factor of angle π to a position number or a position vector in a plane makes the implied direction of the number or vector rotate by 180° with the resultant direction being reversed.

6. Two-dimensional position numbers

As we have discussed before, real numbers form a number line and are one-dimensional numbers. Multiplying a number in the line by another real number makes the corresponding point stretch to another point along the line. On the other hand, multiplying a position number or position vector by a rotation factor makes the corresponding point rotate relative to an origin point. The rotation is no longer a one-dimensional phenomenon that is confined to the point stretching in a straight line, and needs to be viewed from a two-dimensional perspective.

The rotation-generated position number $re^{i\theta}$ represents a point in a two-dimensional plane. In fact, just like real numbers, two-dimensional position numbers magically also possess all of the four basic number properties for all of the four basic arithmetic operations. Despite the similarities, it is also important to identify the differences between the two number types.

6.1 Components of two-dimensional position numbers

First, let's examine the case of one-dimensional line. All real numbers represented by real number set ℝ form a real number line designated as x-axis. Multiplication of a real number by another real number will stretch the corresponding point to reach the other position along the one-dimensional line as illustrated in the figure:

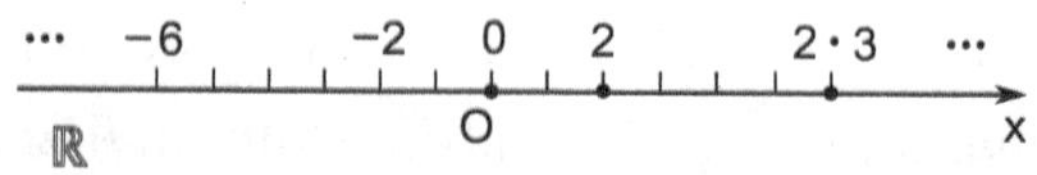

Figure 6.1a. Multiplication of real numbers and position stretching confined in one-dimensional line

Any specific point in the one-dimensional line can be uniquely determined by one parameter, which is a real number value. The parameter may be called coordinate in a coordinate system, or called component in a number or vector system. In other words, from the component viewpoint, real numbers are one-component numbers. Any specific real number has one component, which is the real number itself.

Now, let's look at the case of two-dimensional plane. As shown in this figure,

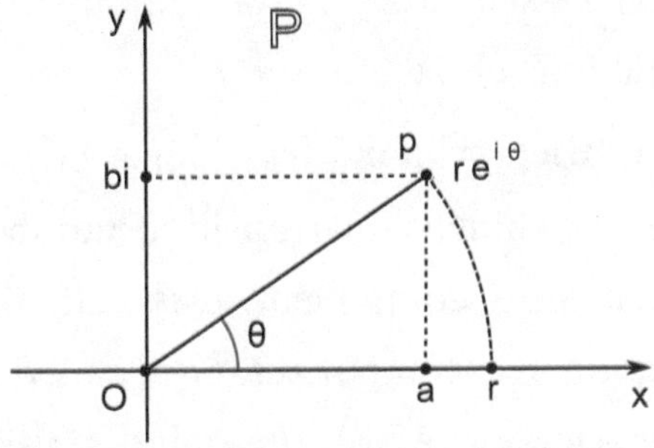

Figure 6.1b. Multiplication of real number r by rotation factor $e^{i\theta}$ and position rotation in two-dimensional plane

all position numbers represented by position number set $\mathbb{P}$ form a two-dimensional plane. Multiplication of a real number r in the x-axis by rotation factor $e^{i\theta}$ rotates the point away from the one-dimensional x-axis and makes it reach the point represented by position number p or $re^{i\theta}$ in the two-dimensional plane.

In contrast to the case of one-dimensional line, any specific point in the two-dimensional plane needs two parameters for the unique determination:

In a polar coordinate system, the two parameters are the magnitude r and the rotation angle θ.

In a Cartesian coordinate system, the two parameters are the x-y coordinates, a and b.

In a vector system, the two parameters are coefficients a and b of the vector's x-y components, $a\hat{\mathbf{u}}_x$ and $b\hat{\mathbf{u}}_y$, where $\hat{\mathbf{u}}_x$ and $\hat{\mathbf{u}}_y$ are respectively x- and y-axis unit vectors.

In a Cartesian position number system, the two parameters are a and b of the position number's x-y components, a and bi, where 1 and i are respectively x- and y-axis unit vectors. The two x-y components are on the same footing for pointing to two directions that are orthogonal to each other.

In a complex number system, which has been shown to be equivalent to the two-dimensional position number system, the terminology is that the x-component, a is called the real number or real part; and the y-component, bi is called the imaginary number or imaginary part. A specific point is also identified by the two parameters a and b.

The naming of real part and imaginary part has often confused learners and given them the false impression that the two parts are different in nature. From the understanding for position numbers, the two parts are symmetric to each other, and represent, respectively, the x-and y-components for two directions that are orthogonal to each other. In fact, the complex number factor $e^{i\theta}$ is often used in solving a practical equation or problem with the solution being in complex number form. Then, either the real part or the imaginary part of the solution may be taken to compare with measurement. That is, the real part and imaginary part are on an equal footing.

It is noted that the position number system may express a physical motion of an object or an abstract position change of a quantity. For the physical motion, position number $\mathrm{p} = \mathrm{r}e^{i\theta}$ indicates the actual position of an object, and its derivative with respect to time, $\frac{dp}{dt}$ indicates the velocity. For the abstract position change, as a concrete example, we may use the position number system to represent a periodic alternative voltage signal v as illustrated in the figure:

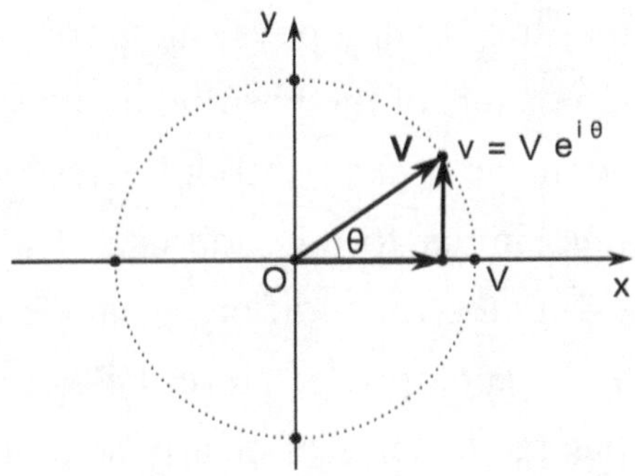

Figure 6.1c. Voltage signal v represented by position number $\mathrm{V}e^{i\theta}$

The point represented by voltage value V in the x-axis is rotated by rotation factor $e^{i\theta}$ to reach the point represented by position number $v = \mathrm{V}e^{i\theta}$, which expresses the voltage at the position. The point represented by the v can also be represented by the voltage vector **V**. When the vector **V** rotates at a constant angular speed ω with the vector's end point v moving in a periodic circular motion, the rotation angle may be expressed as $\theta = \omega t + \phi$, where t is time and ϕ is the initial angle value called phase. We may take either the x-component $\mathrm{V}cos(\omega t + \phi)$ or the-y component $\mathrm{V}sin(\omega t + \phi)$ as the voltage signal for measurement and real-world use. The only difference between the x- and y-components is in the phase shift by $\frac{\pi}{2}$ or 90°. The position number form $\mathrm{V}e^{i\theta}$ has the advantage of achieving a phase shift ϕ_0 by multiplying the corresponding position number $e^{i\phi_0}$: $e^{i\phi_0}\mathrm{V}e^{i(\omega t+\phi)} = \mathrm{V}e^{i(\omega t+\phi+\phi_0)}$. On the contrary, the phase shift cannot be attained with multiplication of a number to the sine or cosine function.

It is also noted that the position number system can be used to represent a quantity (e.g. particle wave function ψ in quantum mechanics) that travels in a physical space. And the naming of x-and y-components of the position number quantity may cause confusion with the x- and y-coordinates in the physical space. In situations similar to this, the use of a more generic name like complex number with different names for the x- and y-components may be appropriate. Nevertheless, the naming of position number has the benefits of clearly stating the nature of the number type, which is the representation of the position for a point in a plane. And the position concept is already familiar to elementary students or beginner learners. It is suited across all dimensions. Real numbers are one-dimensional position numbers. Complex numbers are two-dimensional position numbers. As discussed later, for three and higher n-dimensions, commutative multiplication is still valid for position numbers in spherical coordinate systems. After understanding the concept and nature of the position number, the naming is not that important, and depending on the context, we can easily switch to other relevant names such as complex number if needed.

6.2 Multiplication and division

The 2D position numbers are equivalent to complex numbers. Addition, subtraction, multiplication, and division are applicable to the 2D position numbers along with the associative, commutative, distributive and identity properties. For multiplication and other operations, the property of the orthogonal rotation factor $i^2 = -1$ may be used if needed.

Position numbers $\mathrm{p} = \mathrm{r}e^{i\theta}$ represent all points in a 2D plane. The expression $\mathrm{r}e^{i\theta}$ denotes position numbers in polar coordinates. And the expression $a + bi$ designates position numbers in Cartesian coordinates. Next, we will discuss multiplication and division of position numbers in polar and Cartesian coordinates.

Polar coordinates

For multiplication of position numbers in polar coordinates, let $p_1 = r_1 e^{i\theta_1}$ be one position number, and $p_2 = r_2 e^{i\theta_2}$ be another. The multiplication for the two position numbers is

$$p_1 \cdot p_2 = r_1 r_2 e^{i(\theta_1 + \theta_2)}$$

In terms of direction and magnitude, a position number is equivalent to a vector, and the multiplication can be considered as the operation of using one position number to rotate and stretch the vector of the other position number. Here and throughout, we assume that in the multiplication such as $p_1 \cdot p_2$, the former position number is rotated and stretched by the latter. We can also assume the other way around. The result is the same.

The geometric illustration is shown in the figure:

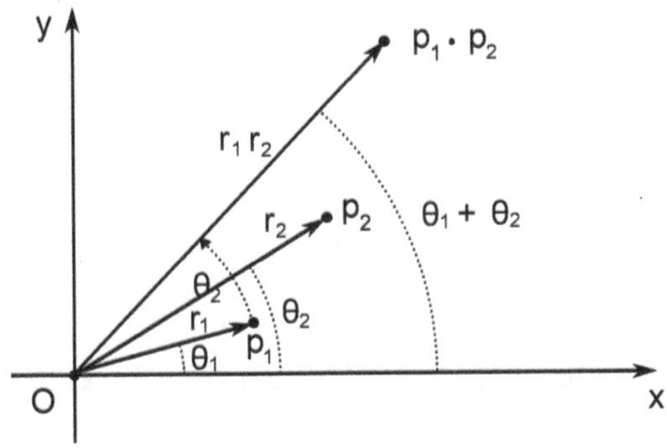

Figure 6.2a. Multiplication of position numbers in polar coordinates

Affected by multiplication of p_2 with angle θ_2 and magnitude r_2, the position number p_1 with angle θ_1 and magnitude r_1, is first rotated by the angle θ_2 with the magnitude r_1 unchanged, and then stretched by the real number r_2 to the position $p_1 \cdot p_2$. The result is the position number $p_1 \cdot p_2$ that has the angle $(\theta_1 + \theta_2)$ and the magnitude $r_1 r_2$.

The multiplication produces a new position number in polar coordinates, and can be transformed by Euler's formula to Cartesian coordinates as

$$p_1 \cdot p_2 = r_1 r_2 \left(cos(\theta_1 + \theta_2) + isin(\theta_1 + \theta_2)\right)$$

The division of position numbers in polar coordinates is given as

$$\frac{p_1}{p_2} = \frac{r_1}{r_2} e^{i(\theta_1 - \theta_2)}$$

After the division by p_2, the position number p_1 is rotated by the angle $-\theta_2$ and stretched by the factor of $\frac{1}{r_2}$ or ${r_2}^{-1}$. The result is a new position number that has the angle $(\theta_1 - \theta_2)$ and the magnitude $\frac{r_1}{r_2}$.

Cartesian coordinates

For multiplication of position numbers in Cartesian coordinates, assume two position numbers as $p_1 = x_1 + y_1 i$ and $p_2 = x_2 + y_2 i$ where x_1, x_2, y_1 and y_2 are real numbers. The product of the two position numbers is

$$
\begin{aligned}
p_1 \cdot p_2 &= (x_1 + y_1 i) \cdot (x_2 + y_2 i) \\
&= x_1x_2 + x_1y_2 i + y_1x_2 i + y_1y_2 i^2 \\
&= x_1x_2 - y_1y_2 + (x_1y_2 + y_1x_2) i
\end{aligned}
$$

Similarly, the multiplication is illustrated in this figure:

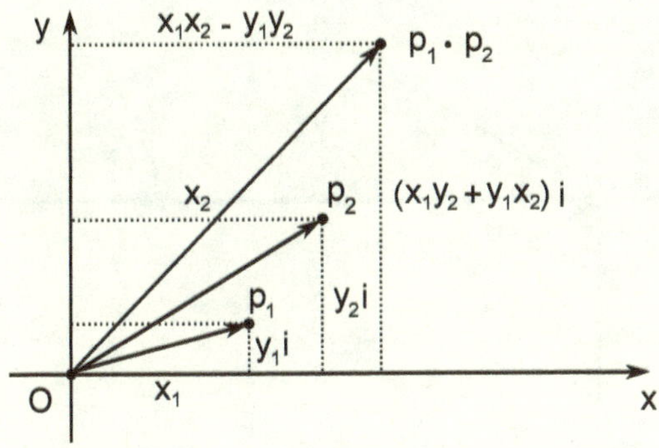

Figure 6.2b. Multiplication of position numbers in Cartesian coordinates

The position numbers are denoted by vectors. The position number p_1 has x-component x_1 and y-component $y_1 i$, and the p_2 has the corresponding x_2 and $y_2 i$. The position number p_1 is rotated and stretched by the position number p_2 through the multiplication. The final result $p_1 \cdot p_2$ is a new position number that has the x-component $(x_1x_2 - y_1y_2)$ and the y-component $(x_1y_2 + y_1x_2)i$.

As an often-used term in mathematics, conjugate of a complex number is introduced. The conjugate of a complex number (position number) $p = x + yi$ is defined as $\bar{p} = x - yi$. It is also written as p^*. The conjugate $\bar{p}$ is the "reflection" of p about the x-axis. In polar form, the conjugate of $re^{i\theta}$ is $re^{-i\theta}$. The product of a position number and its conjugate is a non-negative real number as

$$\mathrm{p} \cdot \bar{\mathrm{p}} = (x + yi)(x - yi) = x^2 + y^2 = r^2$$

The conjugate of a position number and the product are shown in the figure:

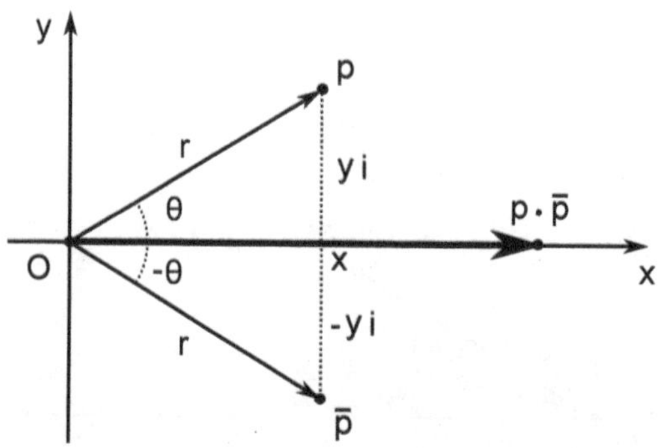

Figure 6.2c. The conjugate $\bar{\mathrm{p}}$ of a position number p and the product $\mathrm{p} \cdot \bar{\mathrm{p}}$

With the conjugate, the multiplicative inverse of a nonzero position number can be computed as

$$\frac{1}{\mathrm{p}} = \frac{\bar{\mathrm{p}}}{\mathrm{p}\bar{\mathrm{p}}} = \frac{x - yi}{x^2 + y^2} = \frac{x}{x^2 + y^2} - \frac{y}{x^2 + y^2} i$$

And with the inverse, the division of two position numbers in Cartesian coordinates can be converted to the multiplication operation as

$$\begin{aligned} \frac{\mathrm{p}_1}{\mathrm{p}_2} &= \mathrm{p}_1 \cdot \frac{1}{\mathrm{p}_2} = (x_1 + y_1 i)\frac{x_2 - y_2 i}{x_2^2 + y_2^2} \\ &= \frac{\mathrm{x}_1\mathrm{x}_2 + \mathrm{y}_1\mathrm{y}_2}{\mathrm{x}_2{}^2 + \mathrm{y}_2{}^2} + \frac{\mathrm{y}_1\mathrm{x}_2 - \mathrm{x}_1\mathrm{y}_2}{\mathrm{x}_2{}^2 + \mathrm{y}_2{}^2} i \end{aligned}$$

6.3 Addition and subtraction

A position number has x-component and y-component whose directions are orthogonal to each other. The addition of two position numbers is computed by respectively adding x- and y-components. That is, the addition is expressed as

$$\mathrm{p}_1 + \mathrm{p}_2 = (x_1 + y_1 i) + (x_2 + y_2 i) = (x_1 + x_2) + (y_1 + y_2)i$$

The geometric representation of the addition is shown in the figure:

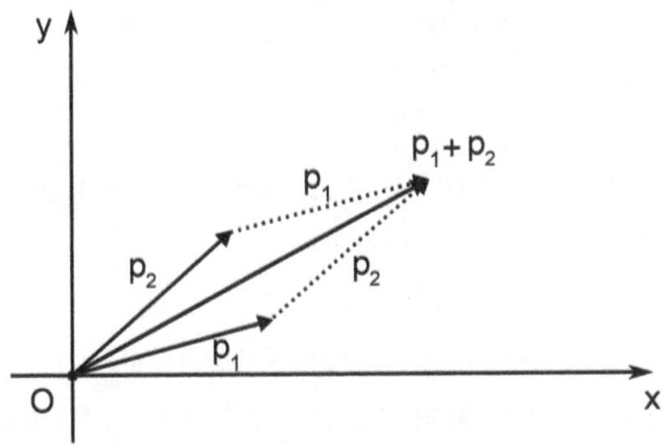

Figure 6.3a. Addition of two position numbers p_1 and p_2

The sum of position numbers p_1 and p_2 is equivalent to the addition of the corresponding vectors. The result is the vector labeled $\mathrm{p}_1 + \mathrm{p}_2$, with the point of the arrow obtained by building a parallelogram from the three vertices: the origin O, and the points of the vector arrows corresponding to the position numbers p_1 and p_2.

Similarly, the subtraction of two position numbers is computed by respectively subtracting x- and y-components. Hence, the subtraction is performed as

$$\mathrm{p}_1 - \mathrm{p}_2 = (x_1 + y_1 i) - (x_2 + y_2 i) = (x_1 - x_2) + (y_1 - y_2)i$$

The geometric representation of the subtraction can be provided by using the fact that subtracting a number is the same as adding its opposite, as illustrated in the figure:

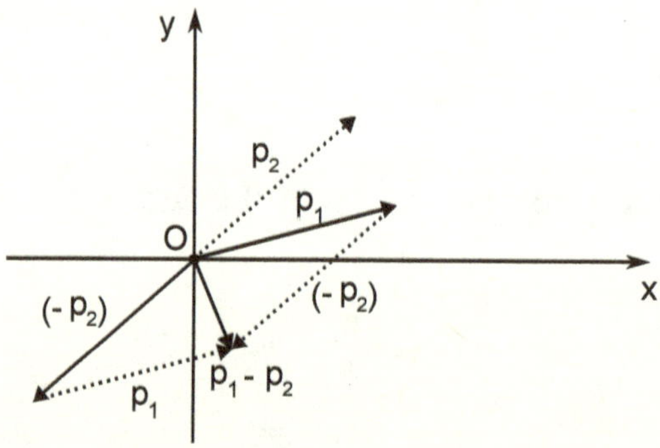

Figure 6.3b. Subtraction of position number p_2 from p_1 being the same as addition of $(-\mathrm{p}_2)$ to p_1

The subtraction of position number p_2 from p_1 is denoted by $\mathrm{p}_1 - \mathrm{p}_2$, which is the same as $\mathrm{p}_1 + (-\mathrm{p}_2)$. In the figure, the vector $(-\mathrm{p}_2)$ is obtained by reversing the direction of the vector p_2. The addition of vector p_1 and vector $(-\mathrm{p}_2)$ produces the resultant vector labeled $\mathrm{p}_1 - \mathrm{p}_2$.

6.4 Vector-complex hybrid algebra

In this section, based on the fact that position vectors can be rotated by rotation factors and the fact that any specific point can be represented by a position vector and a complex number, we will explore a novel concept of vector-complex hybrid algebra where vectors and complex numbers can be mixed and interchangeable for basic arithmetic operations and beyond.

Rotating generic 2D vectors by rotation factors

To be more general, we can include rotation of generic 2D vectors by rotation factors. Previously, we have mainly considered rotation of position vectors relative to the coordinate origin in a plane. In fact, the vector rotation generated by rotation factors is not limited to being relative to the origin, and the target vector can be a generic vector other than a position vector. This is illustrated in the figure:

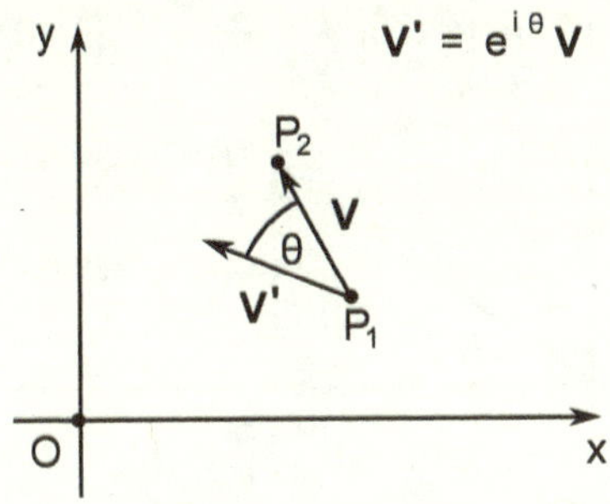

Figure 6.4a. Generic vector rotated by a rotation factor of angle θ

In the figure, vector $\mathbf{V}$ starts from point P_1 and ends at point P_2. With respect to the rotation for vector $\mathbf{V}$ by a rotation factor, it is important to note that the rotation is now relative to the vector starting point P_1, not the origin O. Thus, applying the rotation factor of angle θ to the vector $\mathbf{V}$ makes the vector rotate counterclockwise by angle θ relative to the starting point P_1, and the resultant vector is $\mathbf{V}'$ with the same magnitude. That is, $\mathbf{V}' = e^{i\theta}\mathbf{V}$.

Interchagibility between complex number and vector

Vectors and position numbers are similar to each other. Both have directions and magnitudes. For two-dimensions, position numbers are equivalent to complex numbers. To immediately broaden complex numbers' applications, it is helpful to directly link and integrate position numbers with vectors that are already widely used in physics and engineering.

In the following figure, point P has coordinates (x, y). The point is represented by position vector $\mathbf{P}$ and by position number p.

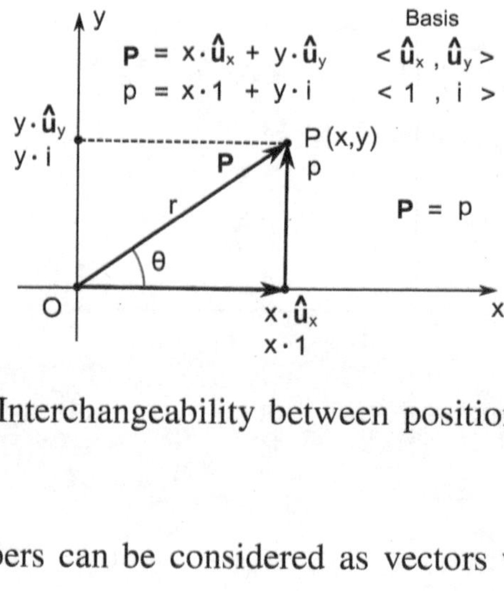

Figure 6.4b. Interchangeability between position numbers and vectors

Position numbers can be considered as vectors with a basis $< 1, i >$. In mathematics, the notation $< 1, i >$ is for a basis, whose elements are unit vectors for the directions of coordinate axes. For position numbers, the basis $< 1, i >$ means that number 1 and orthogonal rotation factor i are unit vectors in x- and y-direction, respectively. For vectors, the basis is $< \hat{\mathbf{u}}_x, \hat{\mathbf{u}}_y >$ with $\hat{\mathbf{u}}_x$ and $\hat{\mathbf{u}}_y$ being unit vectors in x- and y-direction respectively. From the viewpoint of vectors, the unit vectors 1 and $\hat{\mathbf{u}}_x$ are equivalent as they point to the same direction; and the same with the unit vectors i and $\hat{\mathbf{u}}_{\mathrm{y}}$.

With the equivalence of the two bases $< 1, i >$ and $< \hat{\mathbf{u}}_{\mathrm{x}}, \hat{\mathbf{u}}_{\mathrm{y}} >$, we can propose the concept of vector-complex hybrid algebra where a vector and a position number (complex number) are interchangeable with each other if they have the same direction and magnitude. For example, $\mathbf{P}$ and p are interchangeable, and are equal to each other. That is, $\mathbf{P} = \mathrm{p}$. And for $\mathbf{P} = x\hat{\mathbf{u}}_{\mathrm{x}} + y\hat{\mathbf{u}}_{\mathrm{y}}$ and $\mathrm{p} = x + yi$ where x and y are real numbers, the unit vectors $\hat{\mathbf{u}}_{\mathrm{x}}$ and 1 are interchangeable, and the same with the unit vectors $\hat{\mathbf{u}}_{\mathrm{y}}$ and i.

As illustrated in the following equation with $\mathrm{p} = re^{i\theta}$, vector terms and position number terms can be mixed in expressions, and vectors are converted to position numbers for manipulation. Then, the final result is converted back to a vector.

$$
\begin{aligned}
5e^{-i\theta}\mathbf{P} + 3\mathrm{p} + i\mathbf{P} - 2re^{i\theta} &= 5r\hat{\mathbf{u}}_{\mathrm{x}} + 3\mathrm{p} + i\mathrm{p} - 2\mathrm{p} \\
&= 5r + \mathrm{p} + i(x + iy) \\
&= (5r + x - y) + (x + y)i \\
&= (5r + x - y)\hat{\mathbf{u}}_{\mathrm{x}} + (x + y)\hat{\mathbf{u}}_{\mathrm{y}}
\end{aligned}
$$

Another illustration is for obtaining velocity vector $\mathbf{V}$ of circular motion through position number $\mathrm{p} = re^{i\theta}$. The velocity vector is the derivative of the position vector $\mathbf{P}$ with respect to time. The $\mathbf{P}$ is first replaced by p for operations, and then p is changed back to $\mathbf{P}$ as shown:

$$\mathbf{V} = \frac{d\mathbf{P}}{dt} = \frac{d\mathrm{p}}{dt} = ire^{i\theta}\omega = ip\omega = i\mathbf{P}\omega = r\omega\hat{\mathbf{v}}$$

where $\omega = \frac{d\theta}{dt}$ is angular speed, and $\hat{\mathbf{v}}$ is unit vector that is orthogonal to the position vector $\mathbf{P}$. It is noted that $i\mathbf{P}$ means making an orthogonal rotation to the position vector. Therefore, $i\mathbf{P}$ has the magnitude r and the direction $\hat{\mathbf{v}}$.

Multiplication between vector and complex number

The multiplication of vector by rotation factor (complex number in polar form) has been discussed before. Here, we will describe the multiplication of vector and position (complex) number in Cartesian form. As we already know, the orthogonal rotation factor i can rotate vectors including unit vectors $\hat{\mathbf{u}}_\mathrm{x}$ and $\hat{\mathbf{u}}_\mathrm{y}$, with the rotation being 90° counterclockwise. Hence, we have the following identities:

$$\hat{\mathbf{u}}_\mathrm{x} i = i\hat{\mathbf{u}}_\mathrm{x} = \hat{\mathbf{u}}_\mathrm{y} \tag{6.4a}$$

and

$$\hat{\mathbf{u}}_\mathrm{y} i = i\hat{\mathbf{u}}_\mathrm{y} = -\hat{\mathbf{u}}_\mathrm{x} \tag{6.4b}$$

Assume a vector $\mathbf{P}_1$ as $\mathbf{P}_1 = \mathrm{x}_1\hat{\mathbf{u}}_\mathrm{x} + \mathrm{y}_1\hat{\mathbf{u}}_\mathrm{y}$ and a position number p_2 as $\mathrm{p}_2 = \mathrm{x}_2 + \mathrm{y}_2 i$. Then, with the identities above, the multiplication of the vector and the position number is:

$$
\begin{aligned}
\mathbf{P}_1 \mathrm{p}_2 &= (\mathrm{x}_1 \hat{\mathbf{u}}_\mathrm{x} + \mathrm{y}_1 \hat{\mathbf{u}}_\mathrm{y})(\mathrm{x}_2 + \mathrm{y}_2 \imath) \\
&= \mathrm{x}_1 \hat{\mathbf{u}}_\mathrm{x} \mathrm{x}_2 + \mathrm{x}_1 \hat{\mathbf{u}}_\mathrm{x} \mathrm{y}_2 \imath + \mathrm{y}_1 \hat{\mathbf{u}}_\mathrm{y} \mathrm{x}_2 + \mathrm{y}_1 \hat{\mathbf{u}}_\mathrm{y} \mathrm{y}_2 \imath \\
&= (\mathrm{x}_1 \mathrm{x}_2 - \mathrm{y}_1 \mathrm{y}_2) \hat{\mathbf{u}}_\mathrm{x} + (\mathrm{x}_1 \mathrm{y}_2 + \mathrm{y}_1 \mathrm{x}_2) \hat{\mathbf{u}}_\mathrm{y}
\end{aligned}
$$

As we can see, the result is similar to the multiplication between the two corresponding position numbers.

Direct multiplication between vectors

In mathematics, vector multiplication may refer to a specific operation between two vectors. There are different types of vector multiplication such as dot product and cross product. Here, we define a new type of vector multiplication, which may be called direct vector multiplication. Considering the equivalence of a vector and a position number in representing the same point, we may specify the direct multiplication of two vectors as the product of their respective position numbers.

Explicitly, assume a vector $\mathbf{P_1}$ and a position number p_1 representing a point, and another vector $\mathbf{P_2}$ and a position number p_2 representing another point. The direct multiplication $\mathbf{P_1 P_2}$ of the two vectors is defined as

$$
\mathbf{P_1 P_2} = \mathrm{p}_1 \mathrm{p}_2
$$

In polar form, let $\mathrm{p}_1 = \mathrm{r}_1 e^{\imath\theta_1}$ and $\mathrm{p}_2 = \mathrm{r}_2 e^{\imath\theta_2}$. The direct vector multiplication result is

$$
\begin{aligned}
\mathbf{P_1P_2} &= \mathrm{r}_1 e^{i\theta_1}\mathrm{r}_2 e^{i\theta_2} \\
&= \mathrm{r}_1\mathrm{r}_2 e^{i(\theta_1+\theta_2)} \\
&= \mathrm{r}_1\mathrm{r}_2 e^{i(\theta_1+\theta_2)}\hat{\mathbf{u}}_\mathrm{x} \\
&= \mathrm{r}_1\mathrm{r}_2 cos(\theta_1+\theta_2)\hat{\mathbf{u}}_\mathrm{x} + \mathrm{r}_1\mathrm{r}_2 sin(\theta_1+\theta_2)\hat{\mathbf{u}}_\mathrm{y}
\end{aligned}
$$

In other words, the direct vector multiplication rotates and stretches a vector with another vector. The result is a vector with the rotation angle from the x-axis being $\theta_1 + \theta_2$ and the magnitude being $\mathrm{r}_1\mathrm{r}_2$.

And in Cartesian form, let $\mathrm{p}_1 = \mathrm{x}_1 + \mathrm{y}_1 i$ and $\mathrm{p}_2 = \mathrm{x}_2 + \mathrm{y}_2 i$. The direct vector multiplication result is

$$
\begin{aligned}
\mathbf{P_1P_2} &= (\mathrm{x}_1 + \mathrm{y}_1 i)(\mathrm{x}_2 + \mathrm{y}_2 i) \\
&= \mathrm{x}_1\mathrm{x}_2 - \mathrm{y}_1\mathrm{y}_2 + (\mathrm{x}_1\mathrm{y}_2 + \mathrm{y}_1\mathrm{x}_2)i \\
&= (\mathrm{x}_1\mathrm{x}_2 - \mathrm{y}_1\mathrm{y}_2)\hat{\mathbf{u}}_\mathrm{x} + (\mathrm{x}_1\mathrm{y}_2 + \mathrm{y}_1\mathrm{x}_2)\hat{\mathbf{u}}_\mathrm{y}
\end{aligned}
$$

Division between vectors

By the same token as above, we may define division between two vectors as the division of their respective position numbers. That is

$$
\frac{\mathbf{P_1}}{\mathbf{P_2}} = \frac{\mathrm{p}_1}{\mathrm{p}_2}
$$

In polar form, the vector division result is

$$
\begin{aligned}
\frac{\mathbf{P_1}}{\mathbf{P_2}} &= \frac{\mathrm{r}_1 e^{i\theta_1}}{\mathrm{r}_2 e^{i\theta_2}} \\
&= \frac{\mathrm{r}_1}{\mathrm{r}_2} e^{i(\theta_1-\theta_2)}
\end{aligned}
$$

As we can see, the vector division compares the two vectors by the magnitude ratio and by the angle difference. The result is a position number with the magnitude $\frac{\mathrm{r}_1}{\mathrm{r}_2}$ and the angle $\theta_1 - \theta_2$. This is consistent with the other vector division definition previously provided in Formula (4.4).

In Cartesian form, the vector division result is

$$
\begin{aligned}
\frac{\mathbf{P_1}}{\mathbf{P_2}} = \frac{\mathrm{p}_1}{\mathrm{p}_2} &= \mathrm{p}_1 \cdot \frac{1}{\mathrm{p}_2} \\
&= (\mathrm{x}_1 + \mathrm{y}_1 i)\frac{\mathrm{x}_2 - \mathrm{y}_2 i}{\mathrm{x}_2{}^2 + \mathrm{y}_2{}^2} \\
&= \frac{\mathrm{x}_1\mathrm{x}_2 + \mathrm{y}_1\mathrm{y}_2}{\mathrm{x}_2{}^2 + \mathrm{y}_2{}^2} + \frac{\mathrm{y}_1\mathrm{x}_2 - \mathrm{x}_1\mathrm{y}_2}{\mathrm{x}_2{}^2 + \mathrm{y}_2{}^2} i
\end{aligned}
$$

6.5 Position number system and yin-yang number sets

All position numbers ($re^{i\theta}$) form position number set $\mathbb{P}$, and all complex numbers ($a + bi$) form complex number set $\mathbb{C}$. All points in a two-dimensional plane can be represented by the set $\mathbb{P}$ or the set $\mathbb{C}$. Position numbers are equivalent to complex numbers. That is, $\mathbb{P} = \mathbb{C}$. Each member in $\mathbb{P}$ has a corresponding equivalent member in $\mathbb{C}$. For one-dimension, the real number set fulfills the continuous real number line and is considered complete. By the same token, for two-dimensions, the position number set fulfills the continuous number plane and is also considered complete.

The following Table 6.5a summarizes yin-yang number subsets for the position (complex) number set, which is the unity-set. Real number subset for stretching and rotation factor subset for rotation complement each other. It is the contrary and complementary characteristics of the rotation factor subset versus real number subset (or imaginary number subset versus real number subset) that form the complete unity set of position numbers (complex numbers), which in turn form the continuous number plane called position number or complex number plane.

Unity set	**Yin subset •**	**Yang subset ○**
Position numbers ($\mathbb{P}$) / Complex numbers ($\mathbb{C}$)	Rotation factors ($\mathbb{E}$) / Imaginary numbers	Real numbers ($\mathbb{R}$)

Table 6.5a. Yin-yang number subsets forming position/complex numbers

Table 6.5b next consolidates yin-yang number subsets in the complex number system for easy reference and overview.

Unity set	Yin subset •	Yang subset ○
Natural ($\mathbb{N}$)	Zero	Nonzero
Integer ($\mathbb{Z}$)	Negative	Natural
Rational ($\mathbb{Q}$)	Fractional	Integer
Real ($\mathbb{R}$)	Irrational	Rational
Position numbers ($\mathbb{P}$) / Complex numbers ($\mathbb{C}$)	Rotation factors / Imaginary numbers	Real

Table 6.5b. Yin-yang number subsets in the complex number system

Position/complex numbers extend the real number system in one-dimension to two-dimensions, and are blessed with important unique properties and characteristics that have broad applications in mathematics and many scientific areas.

7. Three-dimensional position numbers

Historically, attempts had been made by mathematicians to extend the standard two-dimensional complex numbers to higher dimensions. The extended numbers are categorically known as hypercomplex numbers. One established example of hypercomplex numbers is the quaternion number system introduced by William Rowan Hamilton in 1843, which is in four dimensions and non-commutative. In general, there are restrictions with respect to commutative and associative properties of hypernumbers in higher dimensions. Ferdinand Georg Frobenius proved in 1877 that for a division algebra over the reals to be finite-dimensional and associative, it cannot be three-dimensional (3D), and there are only three such division algebras: real numbers, complex numbers, and quaternions, which have dimensions of 1, 2, and 4 respectively. The consequence is that multiplication (in Cartesian coordinates) is valid for 1D, 2D and 4D, but not 3D. Thus far, there have been no spherical coordinate systems for hypercomplex numbers. The research work of the author here fills the void.

We live in a three-dimensional space. Next, we are going

to focus on investigating into 3D position number system. It is hoped that in the process, we may gain deeper insights into and appreciation more about position numbers and mathematics.

7.1 Position numbers in spherical coordinate system

Previously, the concept of the rotation factors in 2D has been discussed. Based on rotation factors, the 2D polar and Cartesian coordinate systems have been constructed. Now we will extend the similar construction concept to three-dimensional space.

The construction of three-dimensional coordinate system axes is illustrated in the figure:

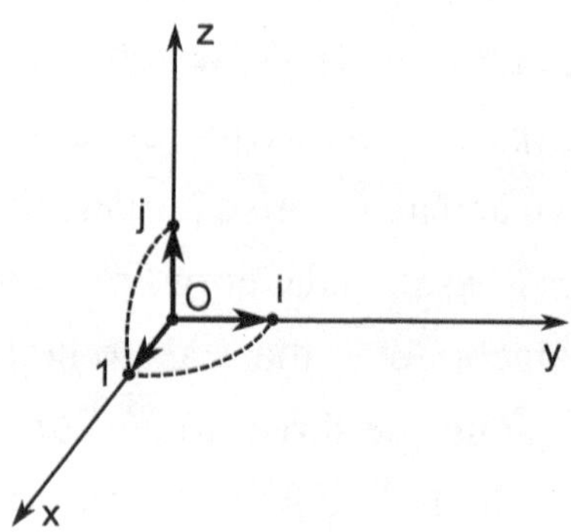

Figure 7.1a. Construction of unit vectors for 3D coordinate system axes

The x-axis is a real number axis. And the y-axis is a real number axis that is orthogonal to the x-axis while the z-axis is a real number axis that is perpendicular to the plane formed by the x-axis and y-axis. The three axes meet at the origin O. Real number 1 is considered as a unit vector pointing to the positive x-axis direction. Applying orthogonal rotation factor i to unit vector 1 makes the vector rotate to unit vector i, which points to the positive y-axis direction. j is another orthogonal rotation factor. Applying the orthogonal rotation factor j to unit vector 1 makes the vector rotate to unit vector j, which points to the positive z-axis direction.

Here, 1, i, and j can be considered as unit vectors for the respective axes, and have been explicitly denoted by arrows. In reality, their directions are implied without the arrow notation. But, in our mind, we can still consider them as unit vectors. On the other hand, i and j are also orthogonal rotation factors that rotate the unit vectors and make them interact with each other. For example, multiplying $-i$ to unit vector i makes the vector rotate back to unit vector 1. Unlike conventional unit vectors, i and j are also rotation factors and can participate in algebra operations, similar to the case for 2D.

After setting up the 3D axes and understanding the characteristics of their unit vectors, we may then proceed to construct the spherical coordinate system, as shown in the figure:

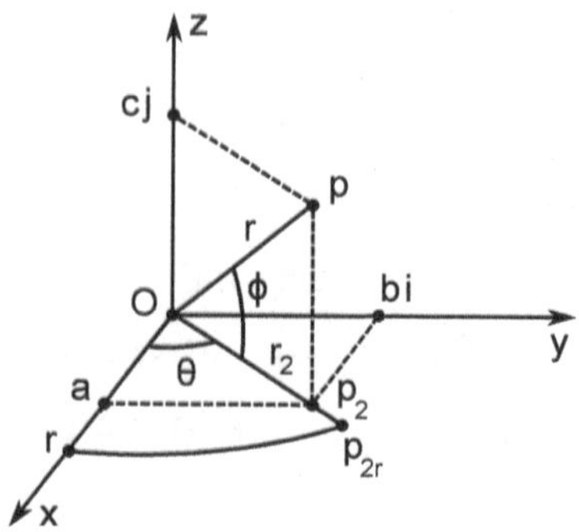

Figure 7.1b. Construction of three-dimensional coordinates by rotation factors

$\mathrm{p_{2r}}$ is a position number representing a point in the second dimension plane that is formed by the x-axis and y-axis. θ is the angle between the position number $\mathrm{p_{2r}}$ and the x-axis. r is a real number in the x-axis. The i-associated rotation factor $e^{i\theta}$ can be used to rotate the number r and make the number rotate to $\mathrm{p_{2r}}$. Thus,

$$\mathrm{p_{2r}} = re^{i\theta}$$

Applying a j-associated rotation factor $e^{j\phi}$ to position number $\mathrm{p_{2r}}$ makes the point rotate by angle ϕ to position number p in the plane $\mathrm{p_{2r}}$-O-z. That is, $\mathrm{p} = \mathrm{p_{2r}}e^{j\phi}$. With the above expression for $\mathrm{p_{2r}}$, the position number p becomes

$$\mathrm{p} = re^{i\theta}e^{j\phi} \tag{7.1}$$

This is the formula of position numbers in the three-dimensional spherical coordinate system. In the formula, r is the magnitude of the position number p; $e^{i\theta}$ the rotation factor for the second dimension plane; and $e^{j\phi}$ the rotation factor for the third dimension in the j or z-axis direction.

As we can see from the formula, the result is succinct and elegant. For a two-dimensional plane, the angle ϕ is zero, and the rotation factor $e^{j\phi}$ is 1. The formula becomes 2D complex numbers in polar coordinates. And for a three-dimensional space, a third dimension is introduced in the j direction and a rotation factor associated with the j direction then appears in the formula.

7.2 Spherical-to-Cartesian transformation and vice versa

With reference to Figure 7.1b, the point represented by position number p_2 is in the plane x-O-y. The line through points p_2 and p is parallel to the z-axis. The position number p_2 has a magnitude of r_2. Like i and j, the unit number 1 may also be considered as a unit vector in the direction of the positive x-axis. In the figure, a, b, and c are real numbers. The terms a, bi, and cj represent, respectively, the components of position number p in the x-, y- and z-directions. In the Cartesian coordinate system, the position number p is expressed as

$$\mathrm{p} = \mathrm{a} + \mathrm{b}i + \mathrm{c}j \tag{7.2a}$$

The geometry in the figure gives

$$\mathrm{a} = rcos(\theta)cos(\phi) \tag{7.2b}$$

$$\mathrm{b} = rsin(\theta)cos(\phi) \tag{7.2c}$$

$$c = r sin(\phi) \tag{7.2d}$$

From (7.2a) through (7.2d), it follows that

$$p = r\left(cos(\theta)cos(\phi) + isin(\theta)cos(\phi) + jsin(\phi)\right) \tag{7.2e}$$

This is the spherical-to-Cartesian coordinate transformation formula.

Likewise, from the geometry or from the above equations (7.2b) through (7.2d), the transformation from Cartesian to spherical coordinates is expressed by

$$r = \sqrt{a^2 + b^2 + c^2} \tag{7.2f}$$

$$\theta = tan^{-1}(\frac{b}{a}) \tag{7.2g}$$

$$\phi = sin^{-1}(\frac{c}{r}) \tag{7.2h}$$

7.3 Commutativity of rotation factors

In the previous Figure 7.1b, the i-associated rotation factor $e^{i\theta}$ is first applied to the number r in the x-axis, and then is followed by the j-associated rotation factor $e^{j\phi}$.

The next Figure 7.3 is the same as Figure 7.1b, except that the order of the operations by the two rotation factors is reversed.

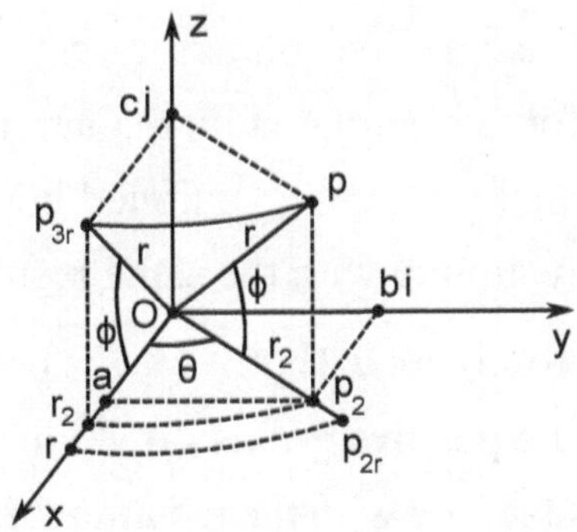

Figure 7.3. Illustration for commutativity of rotation factors

Position number p_{3r} is in the plane x-O-z. The line through p_{3r} and r_2 in the x-axis is parallel to the z-axis. Applying the rotation factor $e^{j\phi}$ to number r in the x-axis makes the point rotate by angle ϕ to position number p_{3r}. And then applying the rotation factor $e^{i\theta}$ to position number p_{3r} makes the point rotate by angle θ to position number p in the plane that is parallel to the plane x-O-y. The final position is the point p, which is the same as the point in Figure 7.1b. That is

$$\mathrm{p} = re^{i\theta}e^{j\phi} = re^{j\phi}e^{i\theta} \tag{7.3}$$

The result (7.3) represents the commutativity of the rotation factors.

7.4 Interaction between orthogonal rotation factors i and j

The orthogonal rotation factors i and j can be considered as unit vectors. On the other hand, they are rotation factors that can rotate vectors. To specify the rotation target, we again use the definition that the former is rotated by the latter. For example, ij means that i is rotated by j. As mentioned before, we may also define the other way around with the same result.

With reference to Figure 7.1b, for ij, i is considered as a unit vector pointing to the positive y-axis direction, and is rotated by orthogonal rotation factor j with the resultant unit vector pointing to the j direction or the positive z-axis direction. That is, ij equals j. And conversely, for ji, j is considered as a unit vector pointing to the z-axis direction, and is rotated by orthogonal rotation factor i. But, the rotation generated by i is in the direction parallel to the plane x-O-y, and cannot make j rotate. That is, ji equals j. We then have this identity:

$$ij = ji = j \tag{7.4a}$$

More generally, any point in the line O-p_{2r} is also rotated to the z-axis by rotation factor j. That is, $e^{i\theta}j = j$. Also, rotation factor $e^{i\theta}$ makes its target vector rotate in the direction parallel to the plane x-O-y, and cannot rotate unit vector j. Thus, $je^{i\theta} = j$. And we have the following identity:

$$e^{i\theta}j = je^{i\theta} = j \tag{7.4b}$$

In fact, the same result can also be obtained by the transformation formula (7.2e). In the spherical coordinate system, rotation factors are commutative. For position number $e^{i\theta}j$ or $je^{i\theta}$, the magnitude r is 1, and the angle ϕ is $\frac{\pi}{2}$. The resultant position number p from the formula is j. That is, we have the identity (7.4b). It should be noted that this rotation factor interaction rule is applicable to spherical coordinate system, but not valid for multiplication of two position numbers in Cartesian coordinate system. For example, the calculation $(1+i+j)i = -1+i+j$ is invalid.

7.5 Coordinate transformation by rotation factor algebra

Previously, we have obtained the spherical-to-Cartesian transformation formula. Next, we will re-produce it by rotation factor multiplication algebra. With Euler's formula for the rotation factor $e^{j\phi}$ in the position number formula $\mathrm{p} = re^{i\theta}e^{j\phi}$, it follows that

$$\mathrm{p} = re^{i\theta}(cos(\phi) + jsin(\phi)) \tag{7.5a}$$

With multiplication, Equation (7.5a) becomes

$$\mathrm{p} = r(e^{i\theta}cos(\phi) + e^{i\theta}jsin(\phi)) \tag{7.5b}$$

And with $e^{i\theta}j = j$ in (7.4b) and Euler's formula for $e^{i\theta}$, Equation (7.5b) leads to

$$\mathrm{p} = r(cos(\theta)cos(\phi) + isin(\theta)cos(\phi) + jsin(\phi)) \tag{7.5c}$$

The result is the same as the formula that is obtained by geometry. It is interesting to see that the spherical-to-Cartesian transformation is also achieved by rotation factor multiplication algebra. This shows the inherent consistency and magic of mathematics.

7.6 Multiplication of position numbers

In the spherical coordinate system represented by the formula $\mathrm{p} = re^{i\theta}e^{j\phi}$, let $\mathrm{p}_1 = \mathrm{r}_1 e^{i\theta_1} e^{j\phi_1}$ be one position number and $\mathrm{p}_2 = \mathrm{r}_2 e^{i\theta_2} e^{j\phi_2}$ be another. Since rotation factors are commutative, the two position numbers are also commutative. The multiplication for the two position numbers is

$$\mathrm{p}_1 \cdot \mathrm{p}_2 = \mathrm{p}_2 \cdot \mathrm{p}_1 = \mathrm{r}_1 \mathrm{r}_2 e^{i(\theta_1+\theta_2)} e^{j(\phi_1+\phi_2)} \tag{7.6a}$$

The result is a new position number in spherical coordinates, and can be transformed by the transformation formula (7.2e) to Cartesian coordinates as

$$\begin{aligned} \mathrm{p}_1 \cdot \mathrm{p}_2 = \mathrm{r}_1 \mathrm{r}_2 (&cos(\theta_1+\theta_2)cos(\phi_1+\phi_2) \\ &+ isin(\theta_1+\theta_2)cos(\phi_1+\phi_2) \\ &+ jsin(\phi_1+\phi_2)) \end{aligned} \tag{7.6b}$$

This formula may be used to circumvent the invalidity of the multiplication in the 3D Cartesian coordinate system. Assume that we have two position numbers, p_1 and p_2 in Cartesian coordinates as $\mathrm{p}_1 = \mathrm{a}_1 + \mathrm{b}_1 i + \mathrm{c}_1 j$ and $\mathrm{p}_2 = \mathrm{a}_2 + \mathrm{b}_2 i + \mathrm{c}_2 j$. For the multiplication of $\mathrm{p}_1 \cdot \mathrm{p}_2$, p_1 and p_2 can be first transformed into spherical coordinates as $\mathrm{p}_1 = \mathrm{r}_1 e^{i\theta_1} e^{j\phi_1}$ and $\mathrm{p}_2 = \mathrm{r}_2 e^{i\theta_2} e^{j\phi_2}$. The multiplication is then performed in the spherical system, and the result for Cartesian coordinates is given by Equation (7.6b).

7.7 Rotating generic 3D vectors

As with the case of 2D, rotation factors can also rotate generic 3D vectors in addition to position vectors. The generic 3D vector rotation formula is expressed as

$$\mathbf{V}' = e^{i\theta} e^{i\phi} \mathbf{V} \tag{7.7}$$

where $\mathbf{V}$ is the target vector that is rotated by two rotation factors, $e^{i\theta}$ and $e^{i\phi}$. The resultant vector is $\mathbf{V}'$.

It is important to note that the rotation angles θ and ϕ are relative to the target vector $\mathbf{V}$ that is to be rotated. This relativeness property of rotation factors is intrinsically fit for describing continuous or successive rotation phenomena such as biological growth, where each current progressive direction is relative to the previous one.

Rotation factors are freely flexible in rotation generations, and encourage their creative novel uses and applications. The rotation formula is concise and elegant, suggesting that it is the native language for the representation of rotations. The vector rotation may find its applications in physics, chemistry (such as describing molecular structures) and other fields.

8. N-dimensional position numbers in spherical coordinates

Before extending results obtained from a three dimensional space to a higher n-dimensional space, we may need to understand the projections from an n-dimensional space to a lower dimensional space. From here on, vectors with bold arrows denote the projections; subscript k (k=1,2,3...n) represents k-dimensional or k-th dimension; and i_k is an orthogonal rotation factor for axis x_k and k-th dimension.

8.1 Projections for n-dimensions

Three-dimensional (3D) space

The projections of position vectors in a 3D space are illustrated in the following figure. Position number p_2 is in the second-dimension plane x_1-O-x_2. The projection of p_2 to the lower first-dimension is the vector corresponding to the component a_1. And the projection to the second-dimension axis x_2 is the component

$a_2 i_2$. That is, $p_2 = a_1 + a_2 i_2$.

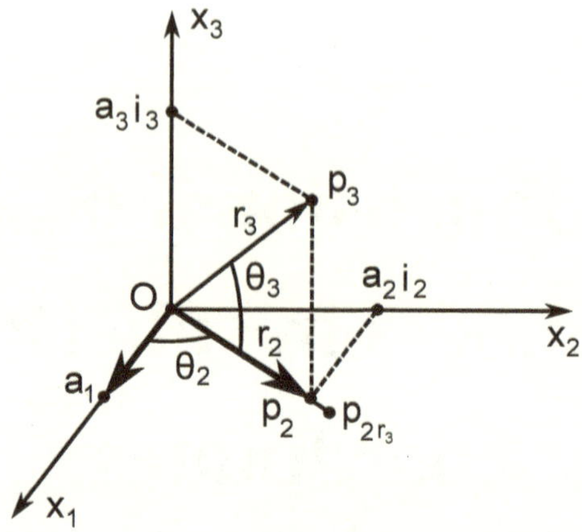

Figure 8.1a. Projections of position vectors in a 3D space

Position number p_{2r_3} is in the same plane as p_2 and also in the same direction. p_{2r_3} has a magnitude of r_3. Applying an i_3-associated rotation factor $e^{i_3\theta_3}$ of angle θ_3 to p_{2r_3} makes the point rotate to position number p_3. The projection of p_3 to the lower second-dimension is p_2. And the projection to the thrid-dimension axis x_3 is the component $a_3 i_3$. The rotation and projections in the figure lead to

$$p_3 = r_3 e^{i_2\theta_2} e^{i_3\theta_3} \tag{8.1a}$$

$$p_3 = p_2 + a_3 i_3 \tag{8.1b}$$

$$r_2 = r_3 cos(\theta_3) \tag{8.1c}$$

$$a_3 = r_3 sin(\theta_3) \tag{8.1d}$$

The 3D equation set of (8.1a)-(8.1d) represents spherical and Cartesian coordinate systems in a 3D space.

Four-dimensional (4D) space

Compared with the 3D space in Figure 8.1a, one more orthogonal axis x_4 is added in the following Figure 8.1b. It is noted that the figure is viewed from the 4D perspective. Logically, if the 4D space occupies our 3D space, which is one dimension lower than the 4D, the 3D space must appear as a 2D plane from the 4D viewpoint. In the 4D space, the x_4 axis is orthogonal to the plane x_2-O-x_3 in the lower dimensions.

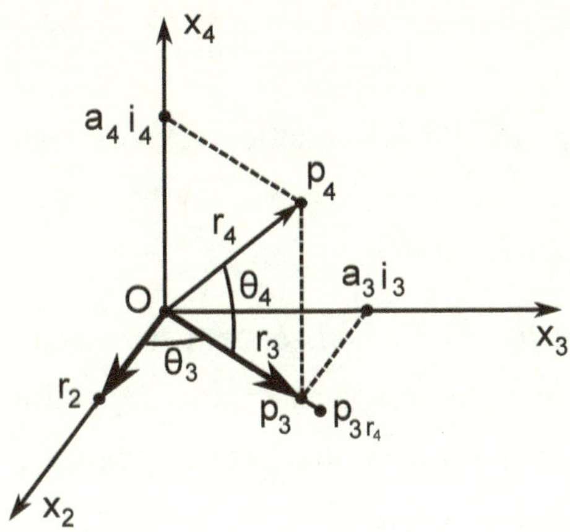

Figure 8.1b. Projections of position vectors in a 4D space

Position number p_{3r_4} is in the same plane as p_3 and also in the same direction. p_{3r_4} has a magnitude of r_4. Therefore, $p_{3r_4}/p_3 = r_4/r_3$ or $p_{3r_4} = r_4p_3/r_3 = r_4e^{i_2\theta_2}e^{i_3\theta_3}$. Applying an i_4-associated rotation factor $e^{i_4\theta_4}$ of angle θ_4 to p_{3r_4} makes the point rotate to position number p_4. Thus, $p_4 = r_4e^{i_2\theta_2}e^{i_3\theta_3}e^{i_4\theta_4}$. The r_4 is the magnitude of p_4. The position vector represented by p_3 in Figure

8.1a now becomes a projection in the current 4D space. The projection of p_4 to the lower 3D dimension is the vector corresponding to p_3. The rotation and projections lead to the corresponding 4D equation set below

$$p_4 = r_4 e^{i_2\theta_2} e^{i_3\theta_3} e^{i_4\theta_4} \tag{8.1e}$$

$$p_4 = p_3 + a_4 i_4 \tag{8.1f}$$

$$r_3 = r_4 cos(\theta_4) \tag{8.1g}$$

$$a_4 = r_4 sin(\theta_4) \tag{8.1h}$$

where p_3 is given by the 3D equation set (8.1a)-(8.1d).

n-dimensional (n-D) space

By the same token, the subsequent figure shows the projections in an n-dimensional space. From the n-dimensional space perspective, the n-1 dimensional space appears as a plane. The x_n axis is orthogonal to the plane x_{n-2}-O- x_{n-1} in the lower dimensions.

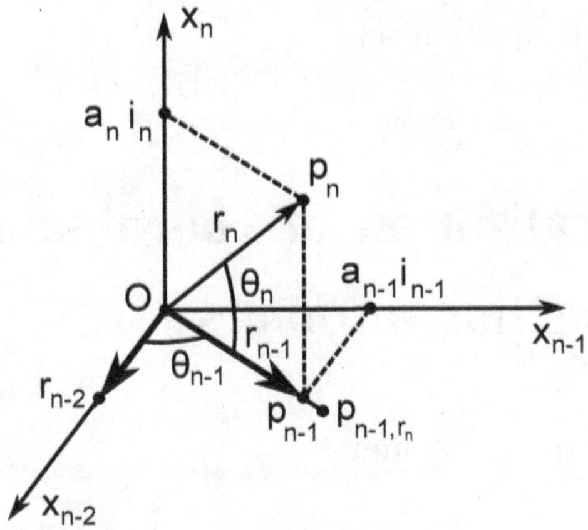

Figure 8.1c. Projections of position vectors in an n-D space

Position number $\mathrm{p}_{n-1,\mathrm{r}_n}$ is in the same plane as p_{n-1} and also in the same direction. $\mathrm{p}_{n-1,\mathrm{r}_n}$ has a magnitude of r_n. Applying an i_n-associated rotation factor $e^{i_n\theta_n}$ of angle θ_n to $\mathrm{p}_{n-1,\mathrm{r}_n}$ makes the point rotate to p_n. The projection of p_n to the lower (n-1)th dimension is p_{n-1}. The above rotation and projections lead to

$$\mathrm{p}_n = \mathrm{r}_n e^{i_2\theta_2} e^{i_3\theta_3} e^{i_4\theta_4} ... e^{i_n\theta_n} \tag{8.1i}$$

$$\mathrm{p}_n = \mathrm{p}_{n-1} + \mathrm{a}_n \mathrm{i}_n \tag{8.1j}$$

$$\mathrm{r}_{n-1} = \mathrm{r}_n cos(\theta_n) \tag{8.1k}$$

$$\mathrm{a}_n = \mathrm{r}_n sin(\theta_n) \tag{8.1l}$$

The n-dimensional equation set of (8.1i)-(8.1l) represents spherical and Cartesian coordinate systems in an n-dimensional space, with p_{n-1} for the lower dimension given by the corresponding n-1 dimensional equation set.

8.2 Generalization of coordinate systems for n-D position numbers

Spherical coordinate system

For the generalization of the coordinate system construction, first in the 2D plane, the 2nd dimension axis x_2 is orthogonal to the 1st dimension axis x_1. The associated rotation factor for the x_2 axis is $e^{i_2\theta_2}$. Then, in the 3D space, the 3rd dimension axis x_3 is introduced to be orthogonal to the plane x_1-O-x_2. The associated rotation factor for the x_3 axis is $e^{i_3\theta_3}$. Similarly, in the n-dimensional space, the n-th dimension axis x_n is introduced to be orthogonal to the plane $\mathrm{x}_{\mathrm{n}-2}$-O-$\mathrm{x}_{\mathrm{n}-1}$. The associated rotation factor for the x_n axis is $e^{i_n\theta_n}$. With (8.1i), the position number in an n-dimensional spherical coordinate system is given by

$$\mathrm{p}_\mathrm{n} = \mathrm{r}_\mathrm{n} \prod_{j=2}^{n} e^{i_j\theta_j} \tag{8.2a}$$

where r_n is the magnitude of the position number and $e^{i_j\theta_j}$ is the j-th dimension associated rotation factor.

Spherical-to-Cartesian transformation

From equation set (8.1b)-(8.1d) for 3D and (8.1f)-(8.1h) for 4D, it follows that

$$\mathrm{p}_3 = \mathrm{r}_3(cos(\theta_2)cos(\theta_3) + i_2 sin(\theta_2)cos(\theta_3) + i_3 sin(\theta_3))$$

and

$$\begin{aligned}\mathrm{p}_4 = \mathrm{r}_4(&cos(\theta_2)cos(\theta_3)cos(\theta_4) \\ &+ i_2 sin(\theta_2)cos(\theta_3)cos(\theta_4) \\ &+ i_3 sin(\theta_3)cos(\theta_4) + i_4 sin(\theta_4))\end{aligned}$$

Along with (8.1j)-(8.1l) for downward iterations, the generalization of n-dimensional position number transformation from spherical coordinates to Cartesian coordinates is given by

$$\mathrm{p}_\mathrm{n} = \sum_{j=1}^{n} a_j i_j \tag{8.2b}$$

where $a_j = r_n E_j$ and

$$E_j = sin(\theta_j) \prod_{k=j+1}^{n} cos(\theta_k)$$

with $i_1 = 1$, $sin(\theta_1) = 1$, and $\prod_{k=j+1}^{n} cos(\theta_k) = 1$ if $j + 1 > n$.

Cartesian-to-spherical transformation

Also from the equation sets (8.1b)-(8.1d), (8.1f)-(8.1h) and (8.1j)-(8.1l), the generalization of n-dimensional position number transformation from Cartesian coordinates to spherical coordinates is given by

$$\mathrm{r_n} = \sqrt{\sum_{j=1}^{n} a_j^2} \tag{8.2c}$$

for the magnitude and

$$(\theta_2, \theta_3, \theta_4, ..., \theta_\mathrm{n}) \tag{8.2d}$$

for (n-1)-tuple of angles with

$$\theta_\mathrm{j} = sin^{-1}(\frac{a_j}{\mathrm{r_j}}) \tag{8.2e}$$

where θ_j is the rotation angle for the j-th dimension with the corresponding magnitude

$$\mathrm{r_j} = \sqrt{\sum_{k=1}^{j} a_k^2} \tag{8.2f}$$

8.3 Generalization of rotation factor algebra for n-D

The rotation factor multiplication algebra involves the interactions of rotation factors and orthogonal rotation factors across different dimensions.

3D space

For the 3D space, with notations in Figure 8.1a, the identity (7.4b) becomes

$$e^{i_2\theta_2} i_3 = i_3 e^{i_2\theta_2} = i_3 \tag{8.3a}$$

And for $\theta_2 = \frac{\pi}{2}$, the above identity becomes

$$i_2 i_3 = i_3 i_2 = i_3 \tag{8.3b}$$

This indicates that the orthogonal rotation factor i_3 is commutative with the orthogonal rotation factor i_2 and the rotation factor $e^{i_2\theta_2}$ in the lower dimension, and the multiplication result is always equal to the i_3 itself.

4D space

In Figure 8.1b for the 4D space, any point in the line O-$\mathrm{p}_{3\mathrm{r}4}$ is rotated to the x_4 axis by orthogonal rotation factor i_4. Thus, $e^{i_3\theta_3} i_4 = i_4$. On the other hand, applying the rotation factor $e^{i_3\theta_3}$ to a point represented by i_4 in the x_4 axis will not rotate the point. Thus, $i_4 e^{i_3\theta_3} = i_4$.

Further, from the 4D perspective, the plane x_1-O-x_2 in Figure 8.1a for the 3D now appears one-dimensional in Figure 8.1b. Logically, in Figure 8.1b, the point represented by the i_2-associated rotation factor $e^{i_2\theta_2}$ is in the x_2 axis. Applying the i_4 orthogonal rotation factor to the point makes it rotate to the x_4 axis. Thus, $e^{i_2\theta_2} i_4 = i_4$. On the other hand, applying the rotation factor $e^{i_2\theta_2}$ to a point represented by i_4 in the x_4 axis will not rotate the point. Thus, $i_4 e^{i_2\theta_2} = i_4$.

From the above, it follows that

$$e^{i_3\theta_3} i_4 = i_4 e^{i_3\theta_3} = e^{i_2\theta_2} i_4 = i_4 e^{i_2\theta_2} = i_4 \tag{8.3c}$$

With angle $\frac{\pi}{2}$ for rotation factors, the above identity leads to

$$i_3 i_4 = i_4 i_3 = i_2 i_4 = i_4 i_2 = i_4 \tag{8.3d}$$

This indicates that the orthogonal rotation factor i_4 is commutative with the orthogonal rotation factors (i_2, i_3) and rotation factors ($e^{i_2\theta_2}$, $e^{i_3\theta_3}$) in the lower dimensions, and the multiplication result is always equal to the i_4 itself.

n-dimensional space

Similarly, in Figure 8.1c for the n-dimensional space, any point in the line O-$\mathrm{p}_{\mathrm{n-1,r_n}}$ is rotated to the $\mathrm{x_n}$ axis by orthogonal rotation factor i_n. Thus, $e^{i_{n-1}\theta_{n-1}}i_n = i_n$. On the other hand, applying the rotation factor $e^{i_{n-1}\theta_{n-1}}$ to a point represented by i_n in the $\mathrm{x_n}$ axis will not rotate the point. Thus, $i_n e^{i_{n-1}\theta_{n-1}} = i_n$.

Further, from the n-dimensional perspective, the point represented by the i_{n-2}-associated rotation factor $e^{i_{n-2}\theta_{n-2}}$ is in the $\mathrm{x_{n-2}}$ axis. Applying the i_n orthogonal rotation factor to the point makes it rotate to the $\mathrm{x_n}$ axis. Thus, $e^{i_{n-2}\theta_{n-2}}i_n = i_n$. On the other hand, applying the rotation factor $e^{i_{n-2}\theta_{n-2}}$ to a point represented by i_n in the $\mathrm{x_n}$ axis will not rotate the point. Thus, $i_n e^{i_{n-2}\theta_{n-2}} = i_n$.

And still further, by the same token, in the n-dimensional coordinate system, all axes are mutually orthogonal to each other. For a j-th dimension with j lower than n-2, applying the i_n orthogonal rotation factor to a point represented by $e^{i_j\theta_j}$ in the j-th dimension makes the point rotate to the $\mathrm{x_n}$ axis. Thus, $e^{i_j\theta_j}i_n = i_n$. On the other hand, applying the rotation factor $e^{i_j\theta_j}$ to a point represented by i_n in the $\mathrm{x_n}$ axis will not rotate the point. Thus,

$i_n e^{i_j\theta_j} = i_n.$

From the above and with the generalization, it follows that

$$e^{i_j\theta_j} i_n = i_n e^{i_j\theta_j} = i_n \tag{8.3e}$$

and

$$i_j i_n = i_n i_j = i_n \tag{8.3f}$$

where $2 \le j < n$.

It is noted that the above results can also be obtained by the spherical-to-Cartesian transformation in (8.2b). Take (8.3e) for example. The position number $e^{i_j\theta_j} i_n = e^{i_j\theta_j} e^{i_n \frac{\pi}{2}}$ means that in (8.2b), $r_n = 1$, $\theta_j = \theta_j$, $\theta_n = \frac{\pi}{2}$ and all other angles (θ_k with $2 \le k \le n-1$ excluding $k = j$) are 0. Thus, all E_j in (8.2b) become 0 except for E_n, which is 1. That is, $\mathrm{p_n} = i_n$, which is consistent with (8.3e).

Obtaining n-dimensional spherical-to-Cartesian transformation by rotation factor algebra

Denote

$$Q_j = \prod_{k=2}^{j} e^{i_k\theta_k} \tag{8.3g}$$

Formula (8.2a) becomes

$$\frac{\mathrm{p_n}}{\mathrm{r_n}} = Q_n \tag{8.3h}$$

The result in (8.3e) means

$$Q_{j-1} i_j = i_j \tag{8.3i}$$

With (8.3i), it follows that

$$\begin{aligned} Q_n &= Q_{n-1} e^{i_n \theta_n} \\ &= Q_{n-1}(cos(\theta_n) + i_n sin(\theta_n)) \\ &= Q_{n-1} cos(\theta_n) + i_n sin(\theta_n) \end{aligned} \tag{8.3j}$$

With (8.3j), it follows that

$$Q_{n-1} = Q_{n-2} cos(\theta_{n-1}) + i_{n-1} sin(\theta_{n-1}) \tag{8.3k}$$

and

$$Q_{n-2} = Q_{n-3} cos(\theta_{n-2}) + i_{n-2} sin(\theta_{n-2}) \tag{8.3l}$$

Q_n may be obtained by continuing the iteration process and combining the iteration results. Inserting the obtained Q_n into (8.3h) leads to Formula (8.2b).

8.4 Comparison of position numbers and hypercomplex numbers

In specific dimensions, the definition of position numbers is different than that of complex or hypercomplex numbers. The former definition is given in spherical coordinates by multiplication of rotation factors and real numbers while the latter is given in Cartesian coordinates by addition of axis components, which are constructed with real numbers for a basis $< 1, i_2, ..., i_n >$.

In two-dimensions, position numbers are shown to be equivalent to complex numbers. However, in other higher dimensions, there are no hypercomplex numbers that are equivalent to position numbers. In a particular dimensions higher than 2D, the interaction between elements in the basis for position numbers is different than that for hypercomplex numbers. Compared with hypercomplex numbers, position numbers should be considered as new number system from the perspective of definition, properties and characteristics.

Number system	2D			3D			4D			8D			n-D		
	MC	MS	CM	MC	MS	CM	MC	MS	CM	MC	MS	CM	MC	MS	CM
Complex number	✓	✓	✓												
Quaternion							✓								
Octonion										✓					
Position number	✓	✓	✓		✓	✓		✓	✓		✓	✓		✓	✓

MC: Multiplicative (Cartesian); MS: Multiplicative (Spherical); CM: Commutative

Table 8.4. Comparison of number systems

The table above shows the comparison of different number systems in 2D and higher dimensions. In two-dimensions, position numbers and complex numbers are equivalent, having the properties of multiplication commutative (CM) for multiplication in Cartesian coordinates (MC) and multiplication in spherical coordinates (MS). In 3D, multiplication in Cartesian coordinates is not valid, but multiplication in spherical coordinates can still be performed for position numbers with commutativity. This is also applicable to any other higher dimensions for position numbers. The conventional rules of multiplication in Cartesian coordinates for hypercomplex numbers such as quaternions (4D) and octonions (8D) are not valid for position numbers in dimensions higher than 2D. However, to achieve the corresponding multiplication result, one may first convert position numbers in Cartesian to spherical coordinates and perform the multiplication in spherical coordinates and transform the result back to Cartesian. The formula for rotation-factor constructed position numbers is succinct and consistent across dimensions with one rotation factor for each dimension, indicating that the rotation factors are natively suited for the spherical systems.

9. Applications of complex numbers in mathematics and science

After understanding two-dimensional position numbers and the equivalence to complex numbers, we will proceed to discuss their applications next. The conventional naming of complex numbers will be used unless stated otherwise.

Complex numbers are important in mathematics, finding applications in a variety of mathematical branches such as algebra and complex analysis. In some areas, they are indispensable in solving problems that cannot be addressed using only real numbers. And in many scientific areas, they are excellent mathematical tools to provide elegant and simplified solutions to problems that appear insurmountable or unmanageable using real numbers alone. The properties and characteristics of the rotation factor $e^{i\theta}$ make it highly useful, particularly the unit circle feature, which is essential for describing and analyzing periodic motions as well as phenomena of vibrations and waves.

9.1 Solutions to quadratic and polynomial equations

Traditionally, the imaginary unit $\sqrt{-1}$ has been introduced as a pure mathematical trick for providing solutions to quadratic and polynomial equations with its real-world connection and causality origin remaining unknown. It is no wonder why this has made many learners feel unsatisfied and unsettled. From the cause-and-effect viewpoint, $\sqrt{-1}$ or $i^2 = -1$ is the effect, not the cause. It is more natural to do the development based on the cause and then obtain the effect instead of the other way around. Historically, the development of the complex number system had been done unconsciously in a top-down approach. Now, the causality origin has been found. It is more intuitive and easier for beginners to study and understand the system in a bottom-up approach.

Quadratic equations

As discussed previously, the position vector rotation by rotation factors is the causality origin for position/complex numbers. Among rotation factors, the orthogonal rotation factor i has the property $i^2 = -1$. This property can provide solutions to quadratic equations in the cases where there are no satisfied answers with real numbers alone. Here, we will use the property and obtain the solutions to the quadratic equation:

$$ax^2 + bx + c = 0$$

where a, b, and c are real numbers with $a \neq 0$.

In the equation, moving the constant c to the right side and dividing the entire equation by a lead to

$$x^2 + \frac{b}{a}x = -\frac{c}{a}$$

Adding $\frac{b^2}{4a^2}$ to both sides of the equation gives

$$x^2 + \frac{b}{a}x + \frac{b^2}{4a^2} = -\frac{c}{a} + \frac{b^2}{4a^2}$$

With rearrangement, we have

$$\left(x + \frac{b}{2a}\right)^2 = \frac{b^2 - 4ac}{4a^2}$$

Let $\Delta = b^2 - 4ac$. The sign of Δ determines if the equation has real number solutions or not. Under the condition $\Delta \geq 0$, the equation has two real number solutions. Next, we will discuss the condition $\Delta < 0$, which gives no real number solutions as the square of any real number is always positive or zero. Under the condition, we have $\Delta = -|\Delta|$ where $|\Delta|$ is the absolute value of Δ. With the notations, the equation can be expressed as

$$\left(x + \frac{b}{2a}\right)^2 = -\frac{|\Delta|}{4a^2}$$

which can be rearranged as

$$\left(\frac{x + \frac{b}{2a}}{\frac{\sqrt{|\Delta|}}{2a}}\right)^2 = -1$$

The equation above becomes $z^2 = -1$ if we let variable z be the term inside the brackets in the above equation. The property $i^2 = -1$ of the orthogonal rotation factor gives solutions $\pm i$ for the equation $z^2 = -1$. Thus, we have

$$\frac{x + \frac{b}{2a}}{\frac{\sqrt{|\Delta|}}{2a}} = \pm i$$

or

$$x = \frac{-b \pm i\sqrt{|\Delta|}}{2a} = \frac{-b \pm i\sqrt{|b^2 - 4ac|}}{2a}$$

Under the condition $b^2 - 4ac < 0$, the quadratic equation has solutions of complex numbers in the form $u + iv$ where u and v are real numbers.

Polynomial equations

A quadratic equation is a polynomial equation of degree two. In fact, complex numbers also provide solutions to polynomial equations of degree n, which are expressed as

$$a_n x^n + ... + a_1 x + a_0 = 0$$

where coefficients a_0, ..., a_n are complex numbers (which also include real numbers).

An equivalent statement of the fundamental theorem of algebra says that every univariate polynomial of positive degree n with complex coefficients can be factorized as $c(x-r_1)...(x-r_n)$, where c, r_1, ..., r_n are complex numbers. This indicates that the n complex numbers r_1, ..., r_n are the roots of the polynomial. If a root appears in several factors, it is a multiple root.

The fact that all roots are in complex numbers makes $\mathbb{C}$ become an algebraically closed field. This property does not apply to the field of rational numbers $\mathbb{Q}$ (e.g. polynomial $x^2 - 2$ has an irrational root $\sqrt{2}$ even though all its coefficients are rational); nor the field of real numbers $\mathbb{R}$ (e.g. polynomial $x^2 + 1$ does not have a real root, in spite of all the coefficients being real numbers).

9.2 Harmonic oscillations and linear differential equations

As we know, complex numbers as position numbers can be expressed as $p = re^{i\theta}$. If the magnitude r remains the same, the position of the point p moves along a circle as the angle θ increases. Now assume that the point moves at a constant angular speed ω. The position is then represented as $p = Ae^{i\omega t}$. The x-projection or x-component of the point is $x = Acos(\omega t)$, which oscillates around zero between $-A$ and A. Here, A is called amplitude. Motion or behavior of this kind is very common and called simple harmonic oscillation.

Simple harmonic oscillator

Simple harmonic oscillator is an example of harmonic oscillation, and is typically illustrated by the motion of an object that is attached to a spring and moves on a flat surface. If the object is displaced a distance x from the equilibrium position 0, the spring will exert an extra force $-kx$ where k is the spring constant.

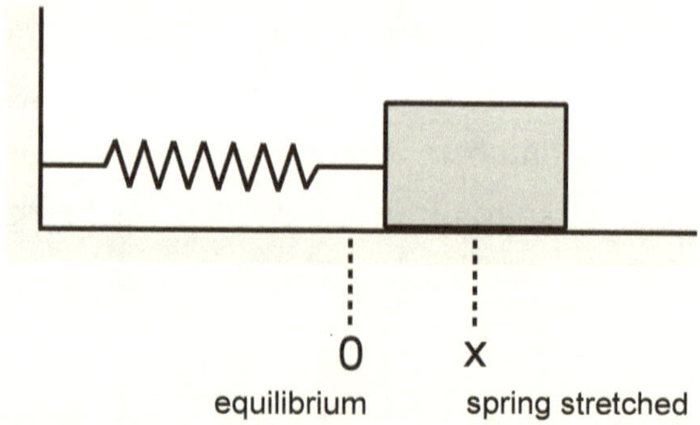

Figure 9.2a. Simple harmonic oscillator

Assume there is no drag force (surface friction and air resistance). The equation of motion will be

$$m\frac{d^2x}{dt^2} = -kx$$

where m is the mass of the object and $\frac{d^2x}{dt^2}$ is the acceleration. The equation is called harmonic oscillator equation, which is a linear differential equation.

The general solution, which can be verified by substitution into the above equation, is

$$x(t) = Acos(\omega_0 t + \phi)$$

where ϕ is called the phase, and A the amplitude. The constant ω_0 is defined as $\omega_0 = \sqrt{k/m}$. The harmonic oscillation is described by a sine or cosine function.

Sinusoidal functions

Sine and cosine functions have a regular period and range, and they have the same shape except for a phase shift of $\frac{\pi}{2}$. This is illustrated in the figure:

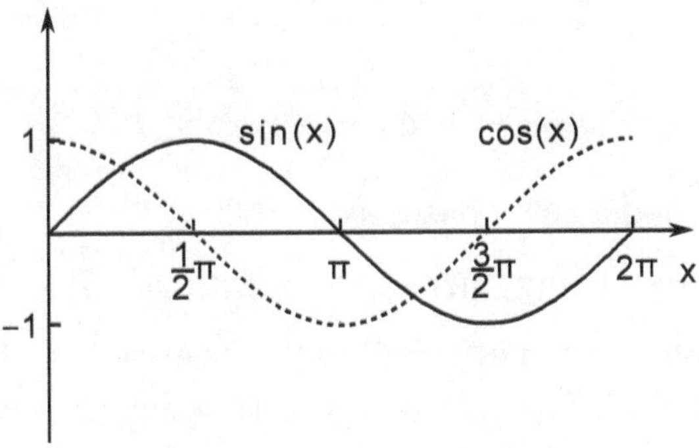

Figure 9.2b. $sin(x)$ and $cos(x)$ having the same shape except for a phase shift of $\pi/2$

A function that has the same general shape as a sine or cosine function is referred to as a sinusoidal function or sinusoid. The general forms of sinusoidal functions are

$$f(x) = Asin(Bx + \phi) + C$$

and

$$f(x) = Acos(Bx + \phi) + C$$

where A, B, and C are constant. ϕ is the phase.

Damped harmonic oscillator

Now assume the motion is damped by a drag force proportional to velocity v of the object: cv, with c being the damping coefficient. We have the damped harmonic oscillator equation as

$$m\frac{d^2x}{dt^2} = -kx - c\frac{dx}{dt}$$

The equation is a linear differential equation, whose general form is

$$a_2\frac{d^2x}{dt^2}+a_1\frac{dx}{dt}+a_0x=0$$

where a_0, a_1 and a_2 are constants.

The equation being linear also means that a linear superposition of two solutions is also a solution. That is, if $x_1(t)$ and $x_2(t)$ are solutions, so is $A_1x_1(t)+A_2x_2(t)$ for any constants A_1, A_2. This can be confirmed by direct substitution into the equation.

The damped harmonic oscillator equation can be re-written as

$$\frac{d^2x}{dt^2}+2\gamma\frac{dx}{dt}+\omega_0^2x=0$$

where $\gamma=\frac{c}{2m}$.

The variable $x(t)$ is the one-dimension displacement in a physical space. However, it is much easier to extend the damped harmonic oscillator equation to complex variable $z(t)$ and solve it. This is due to the fact that the rotation factor is an exponential function and the fact that n-order derivative of an exponential function remains an exponential function, which makes it possible to convert a linear differential equation into an algebraic one. Real numbers may represent a physical dimension or a virtual dimension. Similarly, position/complex numbers may represent a physical plane, a virtual plane, or the combination of x-component in a physical dimension and y-component in a virtual dimension (or vice versa). Here, the x-component denoted by $Re[z(t)]$ represents the real physical x-dimension, while the y-component denoted by $Im[z(t)]$ represents the virtual dimension.

The real number harmonic oscillator equation is extended to the complex one as

$$\frac{d^2z}{dt^2} + 2\gamma\frac{dz}{dt} + \omega_0^2 z = 0 \tag{9.2}$$

With $x = Re[z]$, $\frac{dx}{dt} = Re[\frac{dz}{dt}]$ and $\frac{d^2x}{dt^2} = Re[\frac{d^2z}{dt^2}]$, we have

$$\frac{d^2x}{dt^2} + 2\gamma\frac{dx}{dt} + \omega_0^2 x = Re[\frac{d^2z}{dt^2} + 2\gamma\frac{dz}{dt} + \omega_0^2 z] = 0$$

In other words, a solution $z(t)$ to the complex damped harmonic oscillator equation will also give a solution $x(t)$ to the real damped harmonic oscillator with its real x-component $x(t) = Re[z(t)]$.

Due to the linear superposition, the general solution $z(t)$ may be constructed by two different specific solutions, $z_1(t)$ and $z_2(t)$. That is, $z(t) = A_1 z_1(t) + A_2 z_2(t)$.

To find specific solutions, it is natural to make a guess $z(t) = e^{-i\omega t}$ where ω is a constant to be decided. Inserting the guess into Equation (9.2) gives the conversion from the differential equation to the algebraic one:

$$(-\omega^2 - 2i\gamma\omega + \omega_0^2)e^{-i\omega t} = 0$$

which leads to

$$-\omega^2 - 2i\gamma\omega + \omega_0^2 = 0$$

The two solutions to the quadratic polynomial are

$$\omega_1, \omega_2 = -i\gamma \pm \sqrt{\omega_0^2 - \gamma^2}$$

with ω_1 for the plus signed one and ω_2 for the minus signed.

The linear superposition of the two specific solutions $e^{-i\omega_1 t}$ and $e^{-i\omega_2 t}$ gives the general solution

$$z(t) = A_1 e^{-i\omega_1 t} + A_2 e^{-i\omega_2 t}$$

where A_1 and A_2 are independent free complex number parameters.

Assume $\omega_0 > \gamma$ and define $\omega' = \sqrt{\omega_0^2 - \gamma^2}$. We have

$$\begin{aligned} z(t) &= e^{-\gamma t}(A_1 e^{-i\omega' t} + A_2 e^{i\omega' t}) \\ &= e^{-\gamma t}\left((A_1 + A_2)cos(\omega' t) - i(A_1 - A_2)sin(\omega' t)\right) \end{aligned}$$

The x-component or the real part is the solution to the real damped harmonic oscillator equation:

$$x(t) = Re[z(t)] = e^{-\gamma t}(Acos(\omega' t) + Bsin(\omega' t))$$

where $A = Re[A_1 + A_2]$ and $B = Im[A_1 - A_2]$. The solution can be re-written as

$$x(t) = A' e^{-\gamma t} cos(\omega' t + \phi')$$

where $A' = \sqrt{A^2 + B^2}$ and $\phi' = -tan^{-1}(B/A)$.

The damped harmonic motion is a sinusoidal oscillation with the constant "frequency" ω' (angular speed) and the amplitude modified by a decreasing exponential function $e^{-\gamma t}$, as illustrated in the figure:

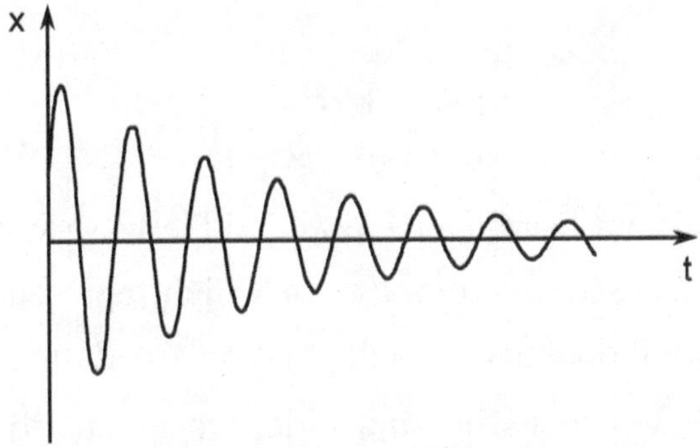

Figure 9.2c. Motion of damped harmonic oscillator

9.3 Fourier series and transform

Fourier series is named after Fourier (1768 - 1830), whose research suggested that an arbitrary continuous (later generalized to any piecewise-smooth) function can be represented by a trigonometric or sinusoidal series. With this great discovery of expressing a function as a sum of sines and cosines, many problems become easier to analyze as the trigonometric functions are well understood. Fourier series were first developed to find the solutions to certain very important partial differential equations, particularly the heat equation and the wave equation.

Waves

Waves are the targets for analysis by Fourier series. In mathematics, physics, engineering, and other related fields, a wave is a propagating perturbation (change from equilibrium) of a certain quantity. Waves are often described by a wave equation, the simplicity form of which is given by the following partial differential equation

$$\frac{\partial^2 s}{\partial x^2} - \frac{1}{v^2}\frac{\partial^2 s}{\partial t^2} = 0$$

where the constant parameter v is called the wave speed, and the variable s is a wave function $s(x, t)$, which represents the value of a perturbation quantity at each position x and time t.

Here, we will focus on sinusoidal waves due to the importance that any perturbation can be described by a sum of sinusoidal oscillations, and any wave can be written as a sum of sinusoidal waves. One family of solutions to the above wave equation are travelling waves of the form

$$s(x, t) = Acos(kx - \omega t + \phi)$$

where $\frac{\omega}{k} = v$. A is called the amplitude of the wave, k the wavenumber, ω the angular speed, and ϕ the phase. The sign of the parameter k, which can be positive or negative, controls the direction of propagation of the wave.

The perturbation value of the wave at a specific point represents the state of the point. A travelling wave means the propagation of each point's state. At a specific time, the state of a specific point is determined by ϕ':

$$\phi' = kx - \omega t + \phi$$

At another time, the state of that specific point becomes the state of another point. In other words, ϕ' is considered constant for the propagation of the state. Taking the partial derivative of ϕ' with respective to time, we have $k\frac{\partial x}{\partial t} - \omega = 0$. That is, $v = \frac{\partial x}{\partial t} = \frac{\omega}{k}$ where v is also called the phase speed of the wave.

The propagation of a travelling wave is illustrated in the following figure. The wave at a specific time (solid curve) has the maximum perturbation state at the point x_0. The wave at another time (dashed curve) propagates by half of the wavelength λ, and the maximum perturbation state travels to the point x'.

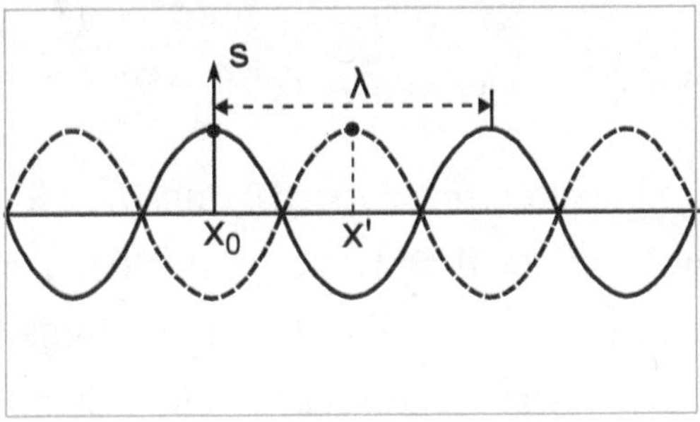

Figure 9.3a. Wave (solid curve) and its propagation (dashed curve)

The required travel time for the distance of the wavelength is the cycle period T. That is, $vT = \lambda$. This leads to $\lambda = \frac{\omega}{|k|}T = \frac{2\pi}{|k|}$. The frequency f of the wave and the angular speed ω has the relationship $\omega = 2\pi f$, with their connections to the cycle period T as $\omega T = 2\pi$ and $f = 1/T$.

By contrast, a pair of superimposed periodic waves traveling in opposite directions produces a standing wave. The wave superpositions are much easier to handle by complex wave representation. In fact, there exist complex solutions to the wave equation. Complex travelling waves, which satisfy the wave equation, have the form

$$s(x,t) = Ae^{i(kx-\omega t+\phi)}$$

The superposition of the two opposite waves of equal amplitude and frequency can be expressed by complex travelling waves as

$$\begin{aligned} s(x,t) &= Ae^{i(kx-\omega t+\phi_1)} + Ae^{i(-kx-\omega t+\phi_2)} \\ &= A\left(e^{i(kx+\phi_1)} + e^{-i(kx-\phi_2)}\right)e^{-i\omega t} \\ &= A\left(e^{i(kx+(\phi_1-\phi_2)/2)} + e^{-i(kx+(\phi_1-\phi_2)/2)}\right)e^{i(\phi_1+\phi_2)/2}e^{-i\omega t} \\ &= 2Acos\left(kx + (\phi_1 - \phi_2)/2\right)e^{i(\phi_1+\phi_2)/2}e^{-i\omega t} \end{aligned}$$

The superposition result represents standing waves.

As an example for standing waves, a string in a music instrument will vibrate at all frequencies that are multiple of its fundamental. Each of these frequencies will have a standing wave on the string. The standing waves for the first five frequencies are illustrated in the figure:

Figure 9.3b. Standing waves of a vibrating string

Fourier series

Fourier series is the sum of harmonics. In physics, music, acoustics, optics, vibration analysis, electrical engineering, signal processing, image processing, telecommunications and other fields, a harmonic is a sinusoidal wave with a frequency that is a positive integer multiple of the fundamental frequency of a periodic signal. The fundamental frequency is also called the 1st harmonic. The other frequencies are known as higher harmonics.

Practically any periodic function that is encountered in applications can be expressed as the sum of harmonics or a Fourier series. For example, a square wave is shown by the graph:

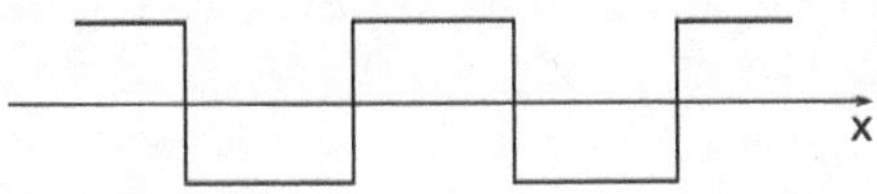

Figure 9.3c. Square wave

The square wave function can be represented by the Fourier series as the sum of harmonics:

$$s(x) = \frac{sinx}{1} + \frac{sin3x}{3} + \frac{sin5x}{5} + \frac{sin7x}{7} + ...$$

In the next figure, the square wave is represented by the first four harmonic terms of the Fourier series. As more harmonic terms are included, their sums converge to the square wave.

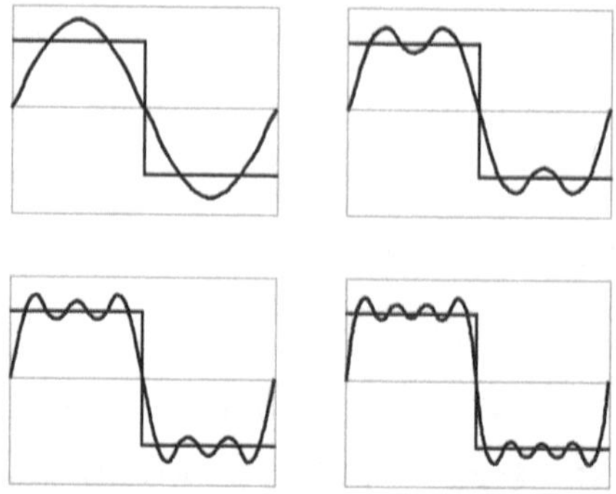

Figure 9.3d. Convergence comparison for square wave represented by Fourier series with 1 term ($\frac{sinx}{1}$) through 4 terms ($\frac{sinx}{1} + ... + \frac{sin7x}{7}$)

A Fourier series is always a continuous, periodic function, but can be used to represent a non-periodic function as shown in the figure:

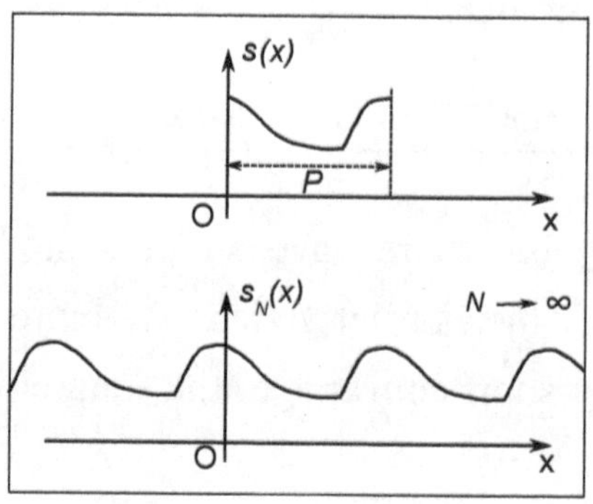

Figure 9.3e. A non-periodic function $s(x)$ in the top graph represented by Fourier series $s_N(x)$ in the bottom graph

The non-periodic function $s(x)$ defined only over the interval from 0 to P in the top graph can be repeated to construct a periodic function with the period of P. The Fourier series $s_N(x)$ for the periodic function is then used to represent the non-periodic function in the said interval.

In general, the Fourier series may be expressed as $s_N(x)$:

$$s_N(x) = a_0 + \sum_{n=1}^{N} \Big(a_n cos(2\pi \frac{n}{P} x) + b_n sin(2\pi \frac{n}{P} x) \Big)$$

The coefficients a_0, a_n and b_n are given by integration over the chosen interval P for the original function $s(x)$ as

$$\begin{aligned} a_0 &= \frac{1}{P} \int_P s(x) dx \\ a_n &= \frac{2}{P} \int_P s(x) cos(2\pi \frac{n}{P} x) dx \\ b_n &= \frac{2}{P} \int_P s(x) sin(2\pi \frac{n}{P} x) dx \end{aligned}$$

As $N \to \infty$, the $s_N(x)$ approaches $s(x)$ at most or all values of x in the interval P.

Don't get intimidated if you are not familiar with integrals. Simply speaking without concerning the rigorousness, an integral is the continuous analog for calculating the area under a curve as an infinite sum of rectangles of infinitesimal width. Notation-wise, the integral of a function $f(x)$ with respect to a variable x on an interval [a, b] is expressed as $\int_a^b f(x)dx$, which means $\lim_{n\to\infty} \sum_{i=1}^{n} f(x_i)\Delta x_i$ where n is the number of equal partitions of the interval, x_i the x value of the i-th partition, and Δx_i the width of the i-th partition.

A Fourier series has other different, but equivalent, forms. In particular, its exponential form is given as

$$s_N(x) = \sum_{n=-N}^{N} c_n e^{i2\pi \frac{n}{P} x}$$

The coefficients c_n are provided by

$$c_n = \frac{1}{P} \int_P s(x) e^{-i2\pi \frac{n}{P} x} dx \tag{9.3}$$

where n belongs to all integers including zero, positive and negative values.

Fourier transform

The exponential form containing complex numbers is most easily generalized for extended applications, which include the finding of frequency information for functions that are not periodic. Fourier series are periodic functions and can be suitable for representing periodic functions. For periodic functions such as the square wave, their Fourier series are formed by harmonics with a set of discrete frequencies. However, for arbitrary functions that are not periodic over the entire real line, their frequency domains can be continuous, not discrete. In this case, Fourier series is not satisfactory and its generalization form called Fourier transform is employed instead. The Fourier transform is an analysis process, in which a function is decomposed into its frequency domain with constituent frequencies and amplitudes. The inverse process is synthesis, in which the original function is recreated from its transform.

The Fourier transform for analyzing a function $s(x)$ into its frequency function $\hat{s}(f)$ is given by

$$\hat{s}(f) = \int_{-\infty}^{\infty} s(x) e^{-i2\pi f x} dx$$

where f is frequency. The result $\hat{s}(f)$ generated by the transform is a complex-valued frequency function, which indicates amplitude and phase of the harmonic at a particular frequency f. The amplitude (magnitude/absolute value) denotes the existence of the particular frequency in the original function. The transform has a zero value for a frequency that does not exist.

The transform formula may be compared with the coefficients of the Fourier series in (9.3), which are indexed by a discrete variable of frequency $\frac{n}{P}$ in the harmonic $e^{-i2\pi\frac{n}{P}x}$. The frequency f is now continuous and so is the function $\hat{s}(f)$, which may be interpreted as the value of the "coefficient" for the harmonic $e^{-i2\pi f x}$ at a particular frequency f.

The following figure illustrates the result of the Fourier transform for a single pulse. The function $s(t)$ of the pulse is in the time domain. The resultant function $\hat{s}(f)$ is in the frequency domain, with the real part (Re) and the imaginary part (Im) denoted by the solid curve and the dashed, respectively.

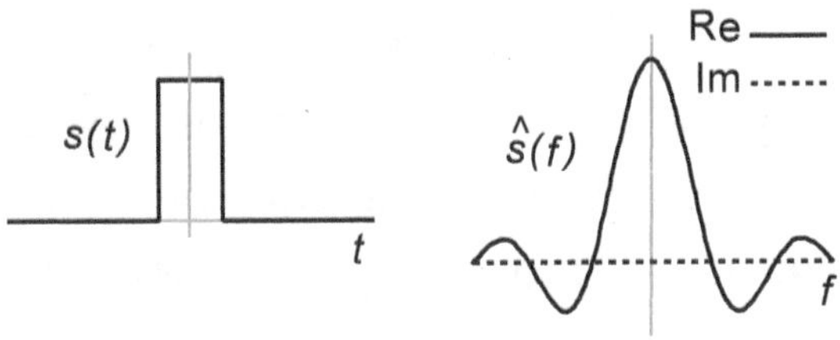

Figure 9.3f. A single pulse and its Fourier transform

The inverse transform is given by

$$s(x) = \int_{-\infty}^{\infty} \hat{s}(f) e^{i2\pi fx} df$$

As an example in analyzing a musical note, the Fourier transform of a sampled musical note would be computed to reveal the frequency components that are present in the note. The same sound could then be re-synthesized by incorporating the frequency components as obtained in the Fourier transform.

The original function's domain is often related to the time domain although the transform is not always restricted to time functions. Functions that are narrowly localized in the time domain will have Fourier transforms that widely spread across the frequency domain, and vice versa. This phenomenon is called the uncertainty principle.

Fourier transform is integral to all modern imaging, which is typically involved with frequencies and certain kinds of waves such as electromagnetic radiation (used in MRI), X-ray (used in CT), and ultrasound, etc. The signal received at the detector (receiver coils in MRI, detector array in CT, and piezoelectric disc in ultrasound) is complicated and made of a large number of constituent frequencies. In the case of spectroscopy for measuring and interpreting electromagnetic spectra, the Fourier transform is used in magnetic resonance imaging (MRI), which is widely employed in hospitals and clinics for medical diagnosis, staging and follow-up of disease. The Fourier transform is also used in nuclear magnetic resonance (NMR), infrared (FTIR) and mass spectrometry.

Fourier analysis

Fourier series, Fourier transform and related extensions such as Fast Fourier Transform (FFT) fall under the umbrella mathematics area called Fourier analysis. Fourier analysis has a vast spectrum of applications - in physics, partial differential equations, number theory, signal processing, digital image processing, probability theory, statistics, forensics, option pricing, cryptography, numerical analysis, acoustics, oceanography, sonar, optics, diffraction, geometry, protein structure analysis, and other areas.

9.4 Electrical circuit analysis

In electrical engineering, electrical circuits serve as the base for developing modern electronic devices and equipments. All electronic computers, instruments, and communication systems require alternating currents and voltages (AC) as well as direct currents and voltages (DC). The notations AC and DC often mean simply alternating and direct, respectively, for specifying the type of either current or voltage or both. AC currents in electrical circuits typically alternate at higher frequencies than those used in power transmission to businesses and residences. Electronic devices may also be used to generate AC by means of a circuit called an oscillator that converts DC into AC.

Alternating currents and voltages of electronic devices are often sinusoidal. And further, any arbitrary signals can be separated into sinusoids through Fourier analysis. Therefore, sinusoidal signals are generally assumed in circuit analysis. The use of complex exponential functions for representing sinusoidal alternating currents and voltages greatly simplifies circuit analysis by converting a differential equation problem into an algebraic one. This makes it possible to express AC equivalents of DC laws such as Ohm's and Kirchhoff's laws. It is noted that in electrical engineering including circuit analysis, the imaginary unit is conventionally denoted by the symbol j instead of i, which is often used to represent electric current.

Phase algebra

Electronic components often cause phase changes to sinusoidal signals. The use of complex numbers is essential in providing the analytic representation for phase change operations.

A real-valued sinusoid with constant amplitude, frequency, and phase has the form:

$$Acos(\omega t + \phi)$$

Multiplication by a constant to the sinusoid cannot change its phase. However, we can use the function $Ae^{j(\omega t+\phi)}$ as an analytic representation of the real-valued sinusoid and achieve the effect through the multiplication by a rotation factor $e^{j\theta}$, which causes a rotation by angle θ and the corresponding change in the phase:

$$\begin{aligned} &Re\left(e^{j\theta}Ae^{j(\omega t+\phi)}\right) \\ =&Re\left(Ae^{j(\omega t+\phi+\theta)}\right) \\ =&Acos(\omega t + \phi + \theta) \end{aligned}$$

The sum of multiple sinusoids with the same frequency produces another sinusoid with an unchanged frequency but a varied phase and amplitude:

$$
\begin{aligned}
&A_1cos(\omega t+\phi_1)+A_2cos(\omega t+\phi_2)\\
=&Re\left(A_1e^{j(\omega t+\phi_1)}+A_2e^{j(\omega t+\phi_2)}\right)\\
=&Re\left((A_1e^{j\phi_1}+A_2e^{j\phi_2})e^{j\omega t}\right)\\
=&Re\left(A_3e^{j\phi_3}e^{j\omega t}\right)\\
=&A_3cos(\omega t+\phi_3)
\end{aligned}
$$

where A_3 and ϕ_3 are, respectively, the amplitude and phase constant of the resultant sinusoid. They may be generally expressed by the parameters A_1, ϕ_1, A_2, and ϕ_2 in long formulas. But, the key point here is that the sum of two complex numbers $A_1e^{j\phi_1}+A_2e^{j\phi_2}$ is another complex number, which can be expressed as its exponential form $A_3e^{j\phi_3}$.

Next, we will discuss the phase difference between voltage and current for each of three basic circuit elements (electronic components): resistor, capacitor, and inductor.

Resistor

A sinusoidal voltage is applied to a resistor (R):

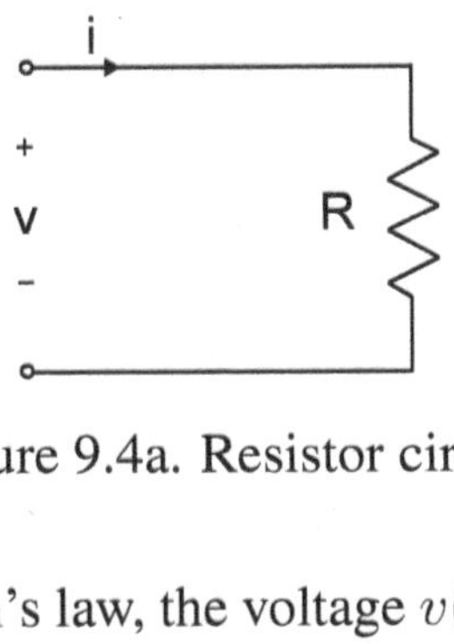

Figure 9.4a. Resistor circuit

According to Ohm's law, the voltage $v(t)$ and the current $i(t)$

of the resistor follows the relation

$$v(t) = i(t)R$$

This indicates that the ratio of the voltage to the current is always constant, regardless of time and signal frequency.

For the voltage signal in the form $v(t) = V_p sin(\omega t)$, we have

$$\frac{v(t)}{i(t)} = \frac{V_p sin(\omega t)}{I_p sin(\omega t)} = R$$

This shows that the voltage and the current are in phase, which means that there is no phase difference between them. The two signals increase and decrease in tandem:

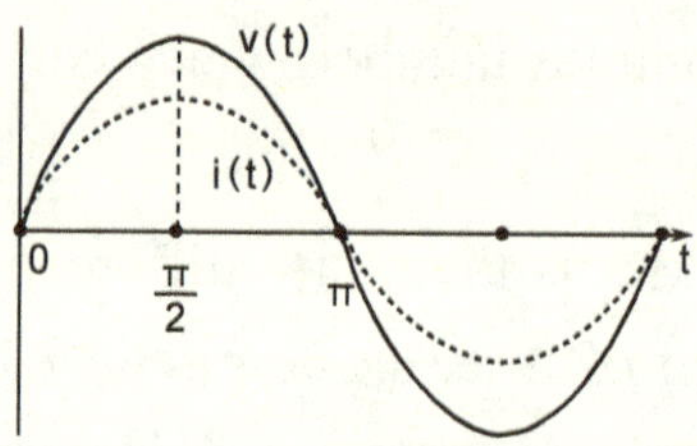

Figure 9.4b. The voltage and current of a resistor being in phase

Capacitor

A capacitor (C) circuit is shown in the figure:

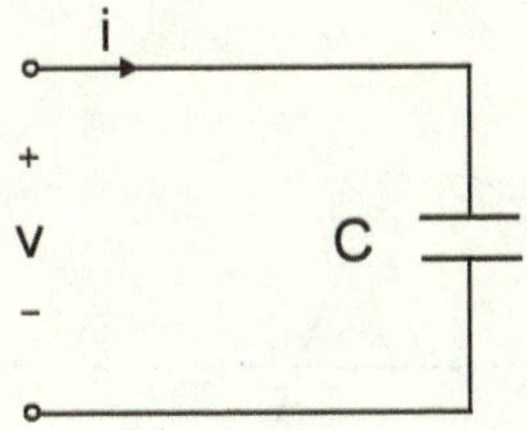

Figure 9.4c. Capacitor circuit

The voltage $v(t)$ across a capacitor and the current $i(t)$ through it follows the relation

$$i(t) = C\frac{dv(t)}{dt}$$

That is, a change in the voltage across a capacitor produces a current that is proportional to the capacitance C and the voltage change rate.

The derivative will change the phase of the signal. Therefore, the voltage signal in the complex representation form $v(t) = V_p e^{j\omega t}$ is assumed. We then have

$$\frac{dv(t)}{dt} = j\omega V_p e^{j\omega t} = j\omega v(t)$$

It follows that

$$\frac{v(t)}{i(t)} = \frac{1}{j\omega C}$$

As we can see, unlike the case for a resistor, the ratio of the voltage to the current is now a complex number, which means the existence of a phase difference. With $\frac{1}{j} = e^{-j\pi/2}$, the negative sign in the angle or phase indicates that the voltage falls behind the current by a phase of $\pi/2$, as illustrated in the following figure:

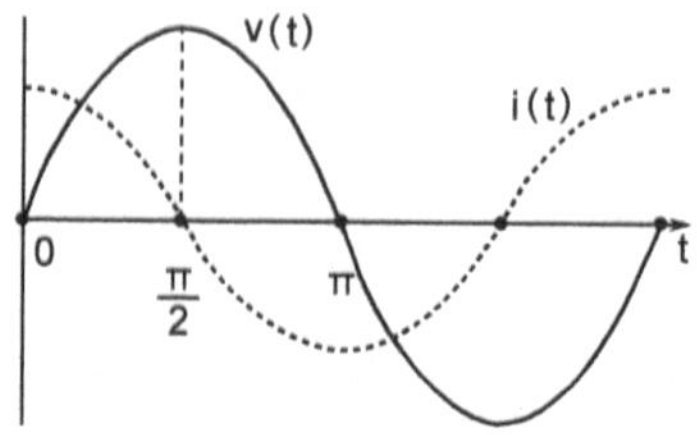

Figure 9.4d. The voltage across a capacitor lagging the current by a phase of $\pi/2$

Inductor

An inductor (L) in a circuit is illustrated in the figure:

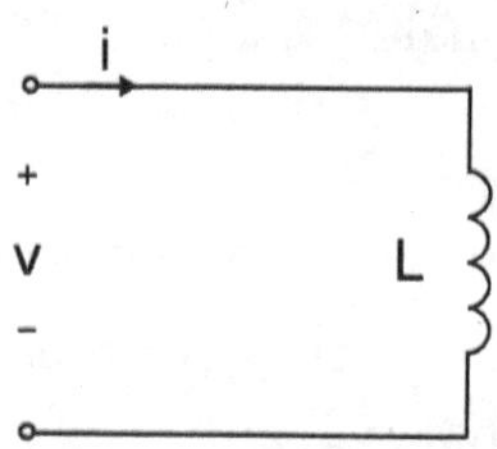

Figure 9.4e. Inductor circuit

The voltage $v(t)$ across an inductor and the current $i(t)$ through it follows the relation

$$v(t) = L\frac{di(t)}{dt}$$

In other words, a change in the current through an inductor produces a voltage across it that is proportional to the inductance L and the current change rate.

Now, the current signal in the complex form $i(t) = I_p e^{j\omega t}$ is assumed. This leads to

$$\frac{di(t)}{dt} = j\omega I_p e^{j\omega t} = j\omega i(t)$$

Hence, we have

$$\frac{v(t)}{i(t)} = j\omega L$$

This tells that for an inductor, the ratio of the voltage to the current is also a complex number. With $j = e^{j\pi/2}$, the positive phase value indicates that the voltage is ahead of the current by a phase of $\pi/2$, which is shown in the figure:

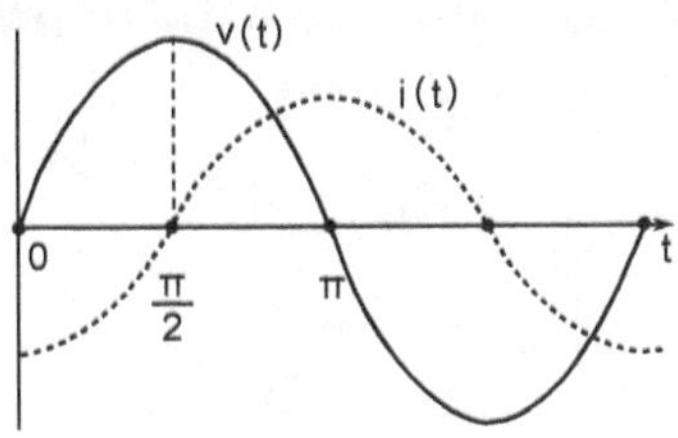

Figure 9.4f. The voltage across an inductor leading the current by a phase of $\pi/2$

Impedance and extending Ohm's law to AC circuits

For the three basic circuit elements, the ratio of voltage to current is a constant, which is known as impedance. Actually, the impedance concept can be extended to other circuit elements beyond the three basic ones.

The impedance Z of a two-terminal circuit element is defined as the ratio of the complex representation of the sinusoidal voltage V between its terminals, to the complex representation of the current I flowing through it. That is

$$Z = \frac{V}{I}$$

In general, impedance is a complex number and can be written as

$$Z = R + jX$$

where the real part of impedance R is called resistance and the imaginary part X is called reactance.

For the three basic elements, we have previously obtained the voltage-to-current ratios, which give the impedances as

$$Z_R = R$$

$$Z_C = \frac{1}{j\omega C}$$

$$Z_L = j\omega L$$

It can be seen that a resistor has the resistance part only, while a capacitor or an inductor has the reactance only. For other circuit elements, their impedances may have both the resistance and the reactance.

In essence, impedance extends Ohm's law to AC circuits. Furthermore, by replacing resistance with impedance, many other rules from DC circuit analysis can also be extended to AC circuits. For example, the rules for combining impedances in series and parallel are identical to those for combining resistances. In contrast, the numbers now are normally complex numbers instead.

For series combination, the current through each circuit element is the same.

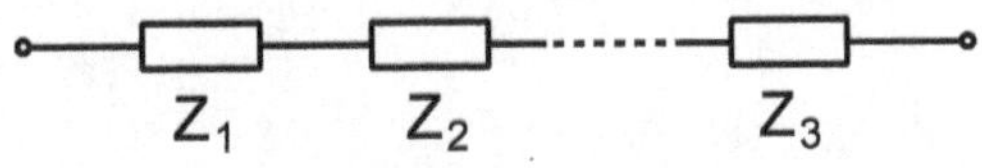

Figure 9.4g. Series combination

The total impedance is therefore the sum of the individual component impedances:

$$Z = Z_1 + Z_2 + ... + Z_n$$

For parallel combination, the voltage across each circuit element is the same.

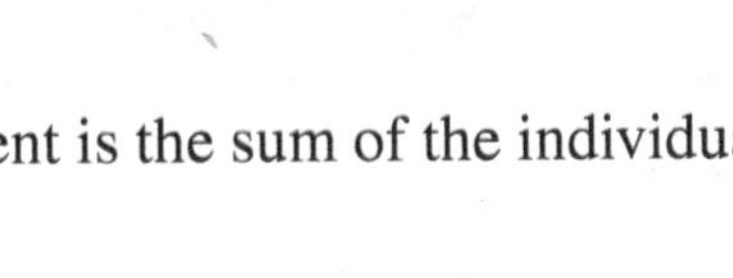

Figure 9.4h. Parallel combination

And the total current is the sum of the individual component currents:

$$I = I_1 + I_2 + ... + I_n$$

or

$$\frac{V}{Z} = \frac{V}{Z_1} + \frac{V}{Z_2} + ... + \frac{V}{Z_n}$$

The result for total impedance Z is

$$\frac{1}{Z} = \frac{1}{Z_1} + \frac{1}{Z_2} + ... + \frac{1}{Z_n}$$

which gives the following for $n = 2$

$$Z = \frac{Z_1 Z_2}{Z_1 + Z_2}$$

LC circuit

An LC circuit, consisting of an inductor and a capacitor, is a resonant circuit for producing signals at a specific frequency or for selecting out a signal at a specific frequency from a more complicated signal. As parts of larger circuits, LC circuits are a key component in various electronic devices for circuits such as oscillators, amplifiers, tuners, filters, and frequency mixers.

The LC circuit serves as an idealized model without the dissipation of energy caused by resistance. There are two types of LC circuit: series LC circuit and parallel LC circuit, depending on the arrangement of the inductor and capacitor.

In a series LC circuit, the inductor and capacitor are connected in series.

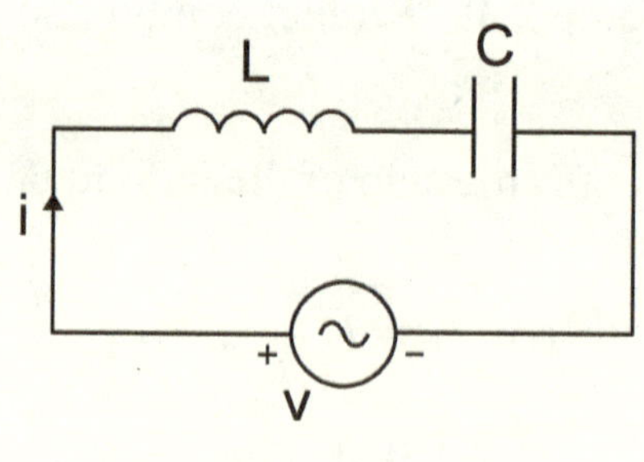

Figure 9.4i. Series LC circuit

The total impedance of the circuit is

$$Z = Z_L + Z_C = j\omega L + \frac{1}{j\omega C} = j(\omega L - \frac{1}{\omega C})$$

At a particular frequency f_0 called the resonant frequency, the two reactance magnitudes $X_L = \omega L$ and $X_C = \frac{1}{\omega C}$ are equal and cancel out each other, leading to the total impedance being zero, with the voltage across the inductor and the voltage across the capacitor being equal in absolute value and opposite in sign. That is, at the resonance, we have

$$\omega L = \frac{1}{\omega C}$$

which gives the solution for ω as

$$\omega = \omega_0 = \frac{1}{\sqrt{LC}}$$

or

$$f_0 = \frac{\omega_0}{2\pi} = \frac{1}{2\pi\sqrt{LC}}$$

In reality, the circuit impedance is not zero due to the existence of non-zero resistance mostly contributed by the resistance of the coil windings. Accordingly, at resonance, the circuit impedance is minimal while the current is maximal.

In a parallel LC circuit, the inductor and capacitor are connected in parallel.

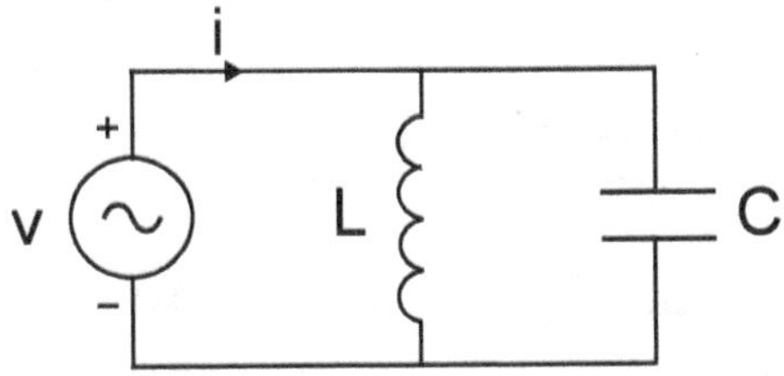

Figure 9.4j. Parallel LC circuit

The total impedance of the circuit is

$$Z = \frac{Z_L Z_C}{Z_L + Z_C} = \frac{j\omega L \frac{1}{j\omega C}}{j\omega L + \frac{1}{j\omega C}}$$

With $\omega_0 = \frac{1}{\sqrt{LC}}$, it can be re-written as

$$Z = -j\frac{1}{C}\left(\frac{\omega}{\omega^2 - \omega_0^2}\right)$$

It can be seen that at the resonance with $\omega = \omega_0$, the impedance becomes infinite. In reality, at resonance, due to the existence of non-zero resistance for the circuit, the impedance is maximal rather than infinite, and the voltage is also maximal. This is in contrast to the series LC circuit where the current is maximal instead.

The LC circuit is referred to as a second-order circuit, meaning that any voltage or current in the circuit can be represented by a second-order differential equation. It is the simplest type of LC network. More complicated and higher order LC networks with more inductors and capacitors generally have more than two reactances and more than one resonant frequency. As we can feel, the use of complex numbers through the impedance provides a powerful means for analyzing sophisticated networks or circuits.

9.5 Quantum mechanics

Quantum mechanics is the field of physics that describes how matter works at and below the scale of atoms. The use of complex numbers is especially essential in quantum mechanics beyond merely for mathematical simplicity and convenience. In fact, the imaginary unit intrinsically appears in the Schrödinger equation, which lies at the heart of quantum mechanics. Measurements show that quantum entities exhibit characteristics of both particles and waves (wave-particle duality). The Schrödinger equation depicts the wave behavior of a quantum particle or the matter wave.

Wave-particle duality of light

The concept of matter wave was originally inspired by the concept of wave-particle duality of light. Max Planck proposed in 1900 that light is quantized into discrete packets of energy. Later in 1905, Albert Einstein conceived that the quantization is inherent to the light with the interpretation of the packets as particles, which are now known as photons. The energy E of a photon is proportional to its frequency f, and given by the Planck-Einstein relation:

$$E = hf$$

where h is known as the Planck constant.

This formula may be expressed in different forms by several characteristic quantities of light, such as frequency f, angular frequency ω, wavelength λ, and wavenumber k, which are related through

$$f = \frac{c}{\lambda} = \frac{\omega}{2\pi} = \frac{ck}{2\pi}$$

where c is the constant for the speed of light. For example, another form is $E = \hbar\omega$, with $\hbar = \frac{h}{2\pi}$.

The relation of the energy and momentum p of a photon is given by Einstein's relativity work as $E = pc$. This combined with the Planck-Einstein relation leads to

$$p = \frac{h}{\lambda} = \hbar k$$

Wave-particle duality of matter

Louis de Broglie in 1924 extended the concept of wave-particle duality to matter, and proposed that the wavelength λ of a particle is related to its momentum p through the the Planck constant h:

$$\lambda = \frac{h}{p}$$

This wavelength is now known as the de Broglie wavelength, and matter waves are also called de Broglie waves. The relations stated for photons, such as $E = \hbar\omega$ and $p = \hbar k$, also hold true for particles.

The momentum p of a non-relativistic particle is $p = mv$, where m is the mass and v the velocity. The associated wavelength is $\lambda = h/mv$. As we can see, the larger the mass, the smaller the wavelength.

The wavelength scale of an electron in contrast to that of a human body is shown in the figure:

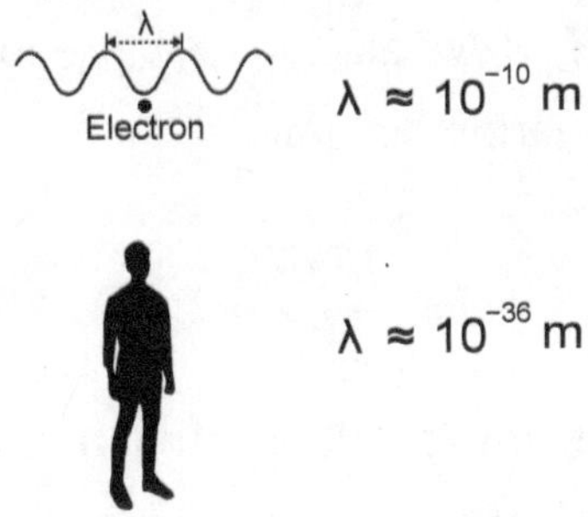

Figure 9.5. Wavelength comparison for an electron and a human body

In reality, the de Broglie wavelength of a matter object is small and difficult to measure. The electron is the lightest particle of matter, and is expected to be the easiest to detect with its relatively long wavelength.

Wave-like behavior of matter has been confirmed by experiments, first for electrons in 1927 and later for other elementary particles, atoms and molecules.

The Schrödinger equation

Based on de Broglie's ideas, in 1926, Erwin Schrödinger published the matter wave equation that is named after him now. Next, we will formulate the Schrödinger equation and see how the imaginary unit comes into the equation natively.

The case of a non-relativistic particle in one dimension is assumed. The classical expression for the energy E of a particle is the sum of the kinetic energy K $(= \frac{1}{2}mv^2 = \frac{p^2}{2m})$ and potential energy V, which is assumed as a function of x only. That is

$$E = K + V = \frac{p^2}{2m} + V(x)$$

With the relations $E = \hbar\omega$ and $p = \hbar k$ proposed by de Broglie, the above energy equation becomes

$$\hbar\omega = \frac{\hbar^2 k^2}{2m} + V(x) \tag{9.5a}$$

As we know, the sinusoidal wavefunction $\psi = cos(kx - \omega t)$ satisfies the classical wave equation. Let $\phi = kx - \omega t$. With $\frac{\partial \psi}{\partial t} = \omega sin\phi$ and $\frac{\partial^2 \psi}{\partial x^2} = -k^2\psi$, we notice that the ω is associated with the first order derivative $\frac{\partial \psi}{\partial t}$, and k^2 is associated with the second order derivative $\frac{\partial^2 \psi}{\partial x^2}$. Therefore, the equation (9.5a) gives a hint of the following form for the wave equation:

$$A\frac{\partial \psi}{\partial t} = B\frac{\partial^2 \psi}{\partial x^2} + V(x)\psi \tag{9.5b}$$

where A and B are constants to be determined.

We now proceed to find a solution to this equation and decide the constants A and B under the condition that the potential energy function $V(x)$ is a constant V_0. The derivative with respect to time in the equation is of the first order rather than the second, as in the classical wave equation. This prevents the wavefunction $\psi = cos(kx - \omega t)$ from being the solution to the matter wave equation. The first order derivative $\frac{\partial \psi}{\partial t}$ produces the $sin\phi$ in contrast to the $cos\phi$ (i.e. ψ itself) that is produced by the second order $\frac{\partial^2 \psi}{\partial t^2}$. This leads to the combination of the sin and cos as a guess for the solution:

$$\psi = cos\phi + \alpha sin\phi$$

where α is a constant to be determined. Inserting this guess and V_0 into the equation (9.5b) results in the following:

$$\frac{A\omega}{\alpha}(-\alpha^2 cos\phi + \alpha sin\phi) = (-Bk^2 + V_0)(cos\phi + \alpha sin\phi) \quad (9.5c)$$

In order to maintain the equality on the left side and right side of the equation for all ϕ values, the term $-\alpha^2$ must be 1. That is, $\alpha^2 = -1$ or $\alpha = \pm i$. We will take the option $\alpha = i$. Here, we notice that the α happens to take the value of the imaginary unit i, which leads to $\psi = cos\phi + \alpha sin\phi = e^{i\phi}$. The reason behind this is that the derivative of $e^{i\phi}$ returns itself, regardless of the order of the derivative. As a result, the derivatives of the term $e^{i\phi}$ cancel out each other in a linear differential equation. With $\alpha = i$, the equation (9.5c) becomes

$$-iA\omega = -Bk^2 + V_0$$

Comparing this equation with $\hbar\omega = \frac{\hbar^2 k^2}{2m} + V_0$ from (9.5a), we have $-iA = \hbar$ and $-B = \frac{\hbar^2}{2m}$. Thus, $A = i\hbar$ and $B = \frac{-\hbar^2}{2m}$. It follows from (9.5b) that

$$i\hbar\frac{\partial\psi}{\partial t} = \frac{-\hbar^2}{2m}\frac{\partial^2\psi}{\partial x^2} + V(x)\psi$$

This is the Schrödinger equation for one dimension. Its 3D form is obtained by using the dependence on all three coordinates (x, y, z) instead of x only, and replacing the operator $\frac{\partial^2}{\partial x^2}$ with $\nabla^2 = \frac{\partial^2}{\partial x^2} + \frac{\partial^2}{\partial y^2} + \frac{\partial^2}{\partial z^2}$.

The formulation above does not prove the validity of the Schrödinger equation, which is a postulate. Actually, it is impossible to prove anything in physics. Nonetheless, we can make an educated guess for a theory, and then test for consistency and agreement with experiments.

Due to the existence of the imaginary unit in the equation, the wavefunction ψ at each point and time is a complex number. The Schrödinger equation describes the behavior of ψ, but states nothing about what it stands for. Shortly after Schrödinger's paper was published, Max Born successfully interpreted ψ as probability amplitude, and $|\psi|^2$ as probability density.

In classical physics, the use of complex numbers is merely a mathematical convenience. In the end, we only take the real part (or the imaginary part, but not both) and can discard the other part. However, in quantum physics, the whole complex wavefunction intrinsically represents the natural behavior of a quantum-mechanical system. Both the real and imaginary parts must be kept. Discarding either part will lose the integrity and produce errors.

Conclusion

By now, after seeing the results obtained from first principles and the real-world connections of imaginary numbers, we may appreciate more about the marvel and charm of mathematics and Nature. If an assumption is based on experience of everyday life, it is much easier to accept along with its consequences. Unlike other main number sets such as natural numbers $\mathbb{N}$, integers $\mathbb{Z}$, rational numbers $\mathbb{Q}$ and real numbers $\mathbb{R}$, complex numbers are based on the assumption of $i = \sqrt{-1}$ or $i^2 = -1$, which is unfamiliar in the physical reality and has hindered its quick adoption by many learners.

From first principles, we have re-created the imaginary unit i and Euler's formula along with other consequences. Vectors are quantities that have both magnitude and direction. They are ubiquitous in their connection to everyday life, encompassing position, displacement, velocity, acceleration, force, momentum and other mechanical or non-mechanical quantities. Vectors can be stretched by real numbers through multiplication. By analogy, the basic assumption that vectors can be rotated by rotation factors through multiplication has been made. Based on the assumption, important consequences or results have been achieved.

With the concept of angle-dependent rotation factors, the fundamental equation of position vector rotation has been discovered. The equation has two important characteristics. One is that an orthogonal rotation factor happens to natively appear in the fundamental equation, which finally reveals the causality origin of the mysterious imaginary unit. The other is that the derivative of position vector with respect to the angle is the position vector itself multiplied by a constant orthogonal rotation factor i, which causes the rotation factor angle-dependence to take the form of an exponential function. The solution to the fundamental equation gives the rotation factor angle-dependence as $e^{i\theta}$, which is miraculously succinct and elegant, with deep implications.

Euler's formula, which lies at the heart of complex number theory, has been derived from a novel approach by constructing Cartesian and polar coordinate systems based on rotation factors without using and relying on $i = \sqrt{-1}$ or $i^2 = -1$. Accordingly, it has been shown that the orthogonal rotation factor is equivalent to the imaginary unit i. The deep meaning behind Euler's formula turns out to be rotation factors and the existence of the rotation factor set.

Complex numbers were first created for the reason that all real polynomials have solutions. Traditionally, for the development of the complex number theory in general and the Euler's formula in particular, a top-down approach has been taken. Specifically, $i = \sqrt{-1}$ is a consequence, not the root cause. And $e^{i\theta}$ is defined in and tied to Euler's formula, which was originally obtained from power series. Later, further study has indicated that i has the characteristic of a 90 degree rotation, and that $e^{i\theta}$ represents a rotation in polar coordinates. In contrast, the rotation factor concept is a bottom-up approach. Based on rotation factors for vector rotation as the root assumption, the orthogonal rotation factor inherently has the property $i^2 = -1$, and the Euler's formula is the natural consequence of the rotation factor angle-dependence $e^{i\theta}$. The bottom-up approach has also revealed other deep implications and findings such as rotation factors for spherical coordinate systems in 3D and n-D, which remain hidden to the top-down approach. In mathematics, there may be more than one method to solve a problem, derive an equation or develop a system. Each method may provide a different perspective and enrich the understanding of the subject matter. This also happens to the complex number system. It is found that there does exist a more thorough way of developing the system with rotation factors from first principles rather than just relying on the imaginary unit, which is the consequence, not the cause. This again demonstrates the marvelous consistency and magic of mathematics.

From the rotation factor concept and its angle-dependence $e^{i\theta}$, the existence of the rotation factor set has been found. The rotation factor set $\mathbb{E}$ is formed by all rotation factors, with the orthogonal rotation factor i being one of its members. A position number is formed by a real number multiplied by a rotation factor. Like the complex number set $\mathbb{C}$, the position number set $\mathbb{P}$ also represents all points in a plane. The position stretching and position rotation are two basic motion types in the physical world. Nature turns out to be fantastically perfect again. The real number set $\mathbb{R}$ for the stretching and the rotation factor set $\mathbb{E}$ for the rotation complement each other and complete the representation for the two basic motion types as well as the whole plane.

From the cause-and-effect viewpoint, the causality origin of the imaginary unit has been revealed through the fundamental equation of position vector rotation. Like the mathematical constant π as the ratio of a circle's circumference to its diameter, the imaginary unit i can be defined as the ratio of two vector terms $\frac{\partial \mathbf{P}}{\partial \theta}$ and $\mathbf{P}$, which indicates an orthogonal rotation. Both terms $\frac{\partial \mathbf{P}}{\partial \theta}$ and $\mathbf{P}$ can be actually simulated and measured in the physical world. The fact that both fundamental mathematical constants π and i share similarities and connections through circle and rotation further manifests the magic of Nature.

With the understandings gained from the 2D development, the concept of rotation factors for rotating position vectors and numbers is extended to other dimensions, and the constructions of three and higher n-dimensional position number spherical coordinate systems are realized. The rotation factor concept allows us to liberate the factor $e^{i\theta}$ from Euler's formula, and provides deep meanings for its application beyond the pure mathematical formula. For example, the relation $e^{i_j\theta_j} i_n = i_n e^{i_j\theta_j} = i_n$ for the rotation factor interaction across different dimensions is obtained by the inherent interpretation of rotation factors, which cannot be achieved with Euler's formula. Rotation factors are multiplication commutative within and across dimensions. The obtained spherical system's coordinate formula is succinct and consistent across dimensions with one rotation factor for each dimension, indicating that the rotation factors are natively suited for the spherical systems.

The further consequences of the basic assumption are the obtaining and understanding of properties and characteristics of rotation factors. Now, with the introduction of rotation factors, the imaginary unit can be directly connected to the physical world through the equivalence of the orthogonal rotation factor i. A rotation factor itself will not rotate, and requires a target to generate a rotation by multiplication. The orthogonal rotation factor i can participate in algebraic operations, and can act as both a rotation generator and a unit vector. A complex number is interchangeable with its vector form as both have the same direction and magnitude. The hybrid algebra of vectors and complex numbers can be promoted to broaden complex numbers' applications. Vectors can be first converted to complex numbers for mathematical manipulations, and the results can then be converted back to vectors by substituting the x- and y component unit vectors.

The rotation factor set $\mathbb{E}$ is a wonderful present from Nature. There indeed exists a set of rotational numbers for rotation besides a set of real numbers for stretching. Discovering the existence of the rotation factor set and understanding its properties and implications immediately give rise to 2D arithmetic. Through a simple notation such as $[\theta]$ along with the readily familiar position and rotation concepts of the real world, 2D arithmetic immediately empowers elementary students to learn and solve practical problems from a higher dimension perspective than 1D.

Special properties and characteristics of complex numbers such as orthogonal rotation factor property $i^2 = -1$ and exponential form $e^{i\theta}$ for rotation factors, become handy when the time comes for their applications. For solutions to polynomials, the property $i^2 = -1$ is a natural fit for the simple quadratic $x^2 + 1 = 0$, and then extended for higher polynomial equations with complex numbers. And with the rotation factor $e^{i\theta}$ that returns itself after derivative and represents periodic circular motion for describing associated oscillation and wave phenomena, complex numbers become indispensable in many application areas of mathematics, physics, engineering and other sciences as well as in Fourier analysis, which is capable of analyzing practically any arbitrary functions and signals.

The technology and science of our modern world are built on the foundation of mathematics, which depends on numbers. Real numbers and complex numbers are two of the most important number types. Real numbers and rotational numbers (rotation factors) complement each other and form the position/complex number set representing a plane, in contrast to the real number set representing a line. Rotational numbers are blessed with unique properties and characteristics with broad applications. The rotational number set is a great gift from Nature – why not use it to the fullest?

Acknowledgments

What a delight it is to write about demystifying imaginary numbers and share with the general public the admiration for the magic and elegance of mathematics and Nature. I've been in awe about the mathematical elegancy and consistency of complex number system, but displeased with the absence of imaginary unit in the physical world ever since first encountering it in high school. Then, decades later, my research work in 2023 unexpectedly derived the imaginary unit from first principles and revealed the existence of imaginary numbers in daily life. The work posted on social media received dominantly positive feedback, which has inspired the writing of this book. My first acknowledgement along with many thanks is hence to you, dear reader in the public for being the source of stimulation.

I would like to express my gratitude and respect to Professor Norihiko Fukuta for guidance and support as well as multiple disciplinary visioning throughout my Ph.D. and later research in University of Utah; to Professor Rong-jue Wei for guidance and open-mind knowledge perusing during my M.S. study and research in Department of Physics, Nanjing University, China; and to my mathematics teacher Suxin Zhuo for study and interest cultivation in high school. I am thankful to Xiaohuan Lu, a faculty of Department of Mathematics, South China Agricultural University for professional discussions and suggestions. I'm deeply grateful to my father Ping Lu and mother Jianping Deng for their spiritual fostering and encouragement towards my education and growth.

Finally, thank you to my wife, Jianping Zheng, for all your love, understanding and cheering-up, and to my sons Alvin and Javin, and daughter Pansy, for discussions and first-hand education experience feedback, which manifest the need for an innovative book on the subject. I am grateful to and love you all.

Qiujiang Lu

Mountain View, California

Bibliography

Allan Pinkus, Samy Zafrany. 1997. *Fourier Series and Integral Transforms*. 1st edn. Cambridge University Press.

Banner, Adrian. 2007. *The Calculus Lifesaver: All the Tools You Need to Excel at Calculus*. Princeton University Press.

Barbeau, E.J. 2003. *Polynomials*. Springer.

Beckmann, Peter. 1976. *History of Pi*. St. Martin's Press.

Blatner, David. 1999. *The Joy of π*. Walker & Company.

Charles Alexander, Matthew Sadiku. 2004. *Fundamentals of Electric Circuits*. McGraw-Hill.

Conway, John B. 1986. *Functions of One Complex Variable*. Springer.

D. Fleisch, L. Kinnaman. 2015. *A student's guide to waves*. Cambridge University Press.

Elias Stein, Rami Shakarchi. 2003. *Fourier Analysis: An introduction*. Princeton University Press.

Feynman, Richard P. 1977. *The Feynman Lectures on Physics*. Vol. I. Addison-Wesley.

G. Lakoff, R.E. Núñez. 2001. *Where Mathematics Comes From*. Basic Books.

Graham, A.C. 1986. *Yin-Yang and the Nature of Correlative Thinking*. Singapore: The Institute of East Asian Philosophies.

Griffiths, David J. 2004. *Introduction to Quantum Mechanics*. 2nd edn. Prentice Hall.

Havil, J. 2012. *The Irrationals. A Story of the Numbers You Can't Count On*. Princeton University Press.

Hughes, Kevin. 2020. *Introduction to the Theory of Yin-Yang*. Independent.

Irwin, Terence H. 1988. *Aristotle's First Principles*. Oxford University Press.

James, J. F. 2011. *A Student's Guide to Fourier Transforms*. 3rd edn. Cambridge University Press.

James W. Nilsson, Susan A. Riedel. 2007. *Electric Circuits*. 8th edn. Pearson Prentice Hall.

James Ward Brown, Ruel V. Churchill. 1996. *Complex variables and applications*. 6th edn. McGraw-Hill.

Kaplan, R. 2000. *The Nothing that Is: A Natural History of Zero*. Oxford University Press.

Liboff, Richard L. 2002. *Introductory Quantum Mechanics*. Addison-Wesley.

Lu, Qiujiang. 2023a. *Deriving the imaginary unit and Euler's formula from first principles, and discovering the existence of rotation factor set*. github.com. https://github.com/mathwonder/Causality-origin-of-the-mysterious-imaginary-unit.

Lu, Qiujiang. 2023b. *Developing two-, three- and n-dimensional complex number spherical coordinate systems from first principles*. github.com. https://github.com/mathwonder/Developing-2D-3D-and-nD-complex-number-spherical-coordinate-systems-from-first-principles.

Maor, Eli. 1998. *"e": The Story of a Number*. Princeton University Press.

Marc-Olivier Renou, Antonio Acín, Miguel Navascués. 2023. *Quantum Physics Falls Apart without Imaginary Numbers*. Scientific American. https://www.scientificamerican.com/article/quantum-physics-falls-apart-without-imaginary-numbers/.

Mazur, B. 2004. *Imagining Numbers (Particularly the Square Root of Minus Fifteen)*. Springer.

MO. Renou, D. Trillo, M. Weilenmann et al. 2021. *Quantum theory based on real numbers can be experimentally falsi-*

fied. Vol. 600. Nature. https://doi.org/10.1038/s41586-021-04160-4.

Nahin, Paul J. 1998. *An Imaginary Tale: The Story of* $\sqrt{-1}$. Princeton University Press.

Nahin, Paul J. 2006. *Dr. Euler's Fabulous Formula: Cures Many Mathematical Ills*. Princeton University Press.

Penrose, Roger. 2016. *The Road to Reality: A complete guide to the laws of the universe*. reprint edn. Random House.

Raymond A. Serway, John W. Jewett. 2003. *Physics for Scientists and Engineers*. Brooks/Cole.

Richard C. Dorf, James A. Svoboda. 2010. *Introduction to Electric Circuits*. 8th edn. John Wiley and Sons.

Richard P. Feynman, Albert R. Hibbs, Daniel F. Steyer. 2005. *Quantum Mechanics and Path Integrals*. Emended edn. McGraw-Hill.

Stillwell, John. 2010. *Mathematics and Its History*. 3rd edn. Springer.

Stipp, David. 2017. *A Most Elegant Equation: Euler's formula and the beauty of mathematics*. Basic Books.

van der Waerden, B. L. 1985. *A History of Algebra*. Springer Verlag.

Wang, Robin R. *Yinyang (Yin-yang)*. Internet Encyclopedia of Philosophy. https://iep.utm.edu/yinyang/ Online retrieved 2024-07-06.

Wilson, Robin. 2018. *Euler's Pioneering Equation: The most beautiful theorem in mathematics*. Oxford University Press.

Index

About the author

Qiujiang Lu is an independent researcher and software developer in Silicon Valley whose research work in mathematics includes "Deriving the imaginary unit and Euler's formula from first principles, and discovering the existence of rotation factor set," which has received high praise from social media users. He obtained his Ph.D. in Atmospheric Sciences (cloud physics) from University of Utah in the US; M.S. in physics from Nanjing University in China; and B.S. in physics from Ocean University of China.

His past research had involved discovering the mechanism of unique ice crystal growth habit change, an unsolved outstanding problem for more than 50 years in cloud physics: ice crystal varies periodically between plate and column shapes under different environment temperatures and super-saturations. The "thousands' variations" in snowflakes are closely related to this habit change mechanism.

Sometimes, people's career trajectories formed under different environments may take the "thousands' variations" like snowflakes. He has multi-backgrounds and interests in physics, mathematics, ocean, atmosphere, cloud physics, air pollution, and software. His physics background and practical-oriented focus happen to strengthen the belief in physical-based causality origin and first principles with the eventual discovery of imaginary numbers in everyday life. His current interests are in physics, mathematics, software, and their multidisciplinary applications.

www.ingramcontent.com/pod-product-compliance
Lightning Source LLC
LaVergne TN
LVHW091531200926
848483LV00002B/361